BLACK&DECKER®

THE COMPLETE GUIDE TO

FLOOR DÉCOR

Beautiful, Long-lasting Floors You Can Design & Install

Creative Publishing international

CHANHASSEN, MINNESOTA
www.creativepub.com

Creative Publishing international, Inc.
18705 Lake Drive East
Chanhassen, Minnesota 55317
1-800-328-3895
www.creativepub.com

Printed at R. R. Donnelley

Library of Congress Cataloging-in-Publication Data

The Complete guide to floor decor : beautiful, long-lasting floors you can design & install / Edited by Clayton Bennett. Branded by Black & Decker.
p. cm.
Summary: "Includes comprehensive information on installing, repairing and maintaining all of the most common and popular floor types"--Provided by publisher.
ISBN-13: 978-1-58923-332-4 (soft cover)
ISBN-10: 1-58923-332-8 (soft cover)
1. Flooring--Handbooks, manuals, etc. 2. Floors--Maintenance and repair--Handbooks, manuals, etc. I. Bennett, Clayton. II. Title.

TH2525.F48 2007
690'.16--dc22 2007018211

Home Improvement Group
Publisher: Bryan Trandem
Managing Editor: Tracy Stanley
Senior Editor: Mark Johanson
Editor: Jennifer Gehlhar

Creative Director: Michele Lanci-Altomare
Senior Design Manager: Brad Springer
Design Managers: Jon Simpson, Mary Rohl

Director of Photography: Tim Himsel
Lead Photographer: Steve Galvin
Photo Coordinators: Adrianne Truthe and Joanne Wawra
Shop Manager: Randy Austin

Production Managers: Linda Halls, Laura Hokkanen

Page Layout Artist: Danielle Smith
Photographers: Andrea Rugg, Joel Schnell
Shop Help: Glenn Austin, John Webb

The Complete Guide to Floor Décor
Created by: The Editors of Creative Publishing international, Inc., in cooperation with Black & Decker.

NOTICE TO READERS

For safety, use caution, care, and good judgment when following the procedures described in this book. Neither the publisher nor Black & Decker® can assume responsibility for any damage to property or injury to persons as a result of misuse of the information provided.
Consult your local Building Department for information on building permits, codes, and other laws as they apply to your project.

Contents

The Complete Guide to Floor Décor

ART OF THE PACIFIC
modern primitives

Introduction

In every room of your house the floor provides a starting point for decorating and furnishing. It's the foundation of how the room functions and the basis for other design choices. You can give any room a new look by changing its contents or wall treatments—but to transform a room entirely, begin at the bottom.

A floor is more than just a place to put your furniture: it establishes the entire feel of a room, giving it color, texture, and mood. By changing an old floor covering to a new one, especially if you use a different material, you can give a room an entirely new sense of character.

Floor coverings range from traditional favorites, such as hardwoods and ceramic tile, to modern materials like compound laminates and recycled rubber. Other possibilities include renewable materials such as bamboo and cork, eye-catching specialties like glass mosaic tile, and even resilient synthetic sheet material made to look like leather. With today's design in flooring, you have more options than ever.

In the past, some flooring materials were expensive to buy and difficult to install. Thanks to the popularity of do-it-yourself projects, manufacturers have developed products and techniques that are accessible to homeowners of all skill levels. The products featured in this book are readily available. Remember to plan ahead, take your time, and get help if you need it.

Planning Your New Floor

Like any successful remodeling project, replacing your floor covering requires detailed planning and attention to design. Flooring is not separate from the rest of the room; it should fit into the overall design to create a desired effect. A floor can create excitement and become a focal point or it can serve as a background for the rest of the room.

Through careful planning, you can choose flooring that can be used successfully in multiple rooms, or select a pattern or design that is repeated throughout the room or in adjacent rooms.

Keep in mind that your flooring design will last a long time, especially if you install ceramic tile or wood. In most cases, the only way to change the design of your floor is to install a new floor covering.

The information in this section will help you plan and design floors that meet your needs. After looking through the portfolio section for ideas of various floor coverings, you may also want to visit a flooring showroom to find the color, style, and pattern of the material you want to use. Then, following the directions in the main section of this book, you can create a new look and feel for your living space.

In this Chapter:

Planning Overview

Because floors are highly visible, appearance is one of the most important considerations when you choose flooring. Start your search by browsing through magazines and visiting flooring retail stores.

Visual appearance is only one factor, of course. You will have several reasons to choose one flooring material over another; appearance is the most obvious. Other characteristics to keep in mind include cost, ease of installation, durability, and how the flooring feels underfoot.

In a room that is subject to moisture, heavy traffic, or other demanding conditions, consider how different floor coverings will perform under these situations. Some flooring cannot be installed in damp areas, while others can fade or scuff under heavy traffic. Read pages 10 through 15 for help in choosing a floor covering that best fits your needs.

When estimating materials for your project, add 10% to 15% to your total square footage to allow for waste caused by trimming. For some carpet installations, you will need to add even more. Save extra flooring material in case future repairs are needed.

Establish a logical work sequence. Many flooring projects are done as part of a more comprehensive remodeling project. In this case, the flooring should be installed after the walls and ceiling are finished, but before the fixtures are installed. Protect new flooring with heavy paper or tarps when completing your remodeling project.

Measure the area of the project room to calculate the quantity of materials you'll need. Measure the full width and length of the space to determine the overall square footage, then subtract the areas that will not be covered, such as stairways, cabinets, and other permanent fixtures.

Checklist for Planning a Flooring Project ▸

Read the following sections and use this checklist to organize your activities as you start your flooring project.

- ❑ Measure the project area carefully. Be sure to include all nooks and closets, as well as areas under all movable appliances. Calculate the total square footage of the project area.
- ❑ Use your measurements to create a floor plan on graph paper.
- ❑ Sketch pattern options on tracing paper laid over the floor plan to help you visualize what the flooring will look like after you install it.
- ❑ Identify areas where the type of floor covering will change, and choose the best threshold material to use for the transition.
- ❑ Estimate the amount of preparation material needed, including underlayment sheets and floor leveler.
- ❑ Estimate the amount of installation material needed, including the floor covering and other supplies, such as adhesive, grout, thresholds, tackless strips, and screws. Add 5 to 10% to your total square footage to allow for waste caused by trimming. For some carpet installations you will need to add even more. Tip: For help in estimating, go to a building supply center and read the labels on materials and adhesives to determine coverage.
- ❑ Make a list of the tools needed for the job. Locate sources for the tools you will need to buy or rent.
- ❑ Estimate the total cost of the project, including all preparation materials, flooring and installation materials, and tools. For expensive materials, shop around to get the best prices.
- ❑ Check with building supply centers or flooring retail stores for delivery costs. A delivery service is often worth the additional charge.
- ❑ Determine how much demolition you will need to do, and plan for debris removal through your regular garbage collector or a disposal company.
- ❑ Plan for the temporary displacement of furnishings and removable appliances to minimize disruption of your daily routine.

NOTES

Floor Anatomy

A typical wood-frame floor consists of layers that work together to provide the required structural support and desired appearance:

1. At the bottom of the floor are the joists, the 2 × 10 or larger framing members that support the weight of the floor. Joists are typically spaced 16" apart on center.
2. The subfloor is nailed to the joists. Most subfloors installed in the 1970s or later are made of ¾" tongue-and-groove plywood; in older houses, the subfloor often consists of 1"-thick wood planks nailed diagonally across the floor joists.
3. On top of the subfloor, most builders place a ½" plywood underlayment. Some flooring materials, especially ceramic tile, require cementboard for stability.
4. For many types of floor coverings, adhesive or mortar is spread on the underlayment before the floor covering is installed. Carpet rolls generally require tackless strips and cushioned padding.
5. Other materials, such as snap-fit laminate planks or carpet squares, can be installed directly on the underlayment with little or no adhesive.

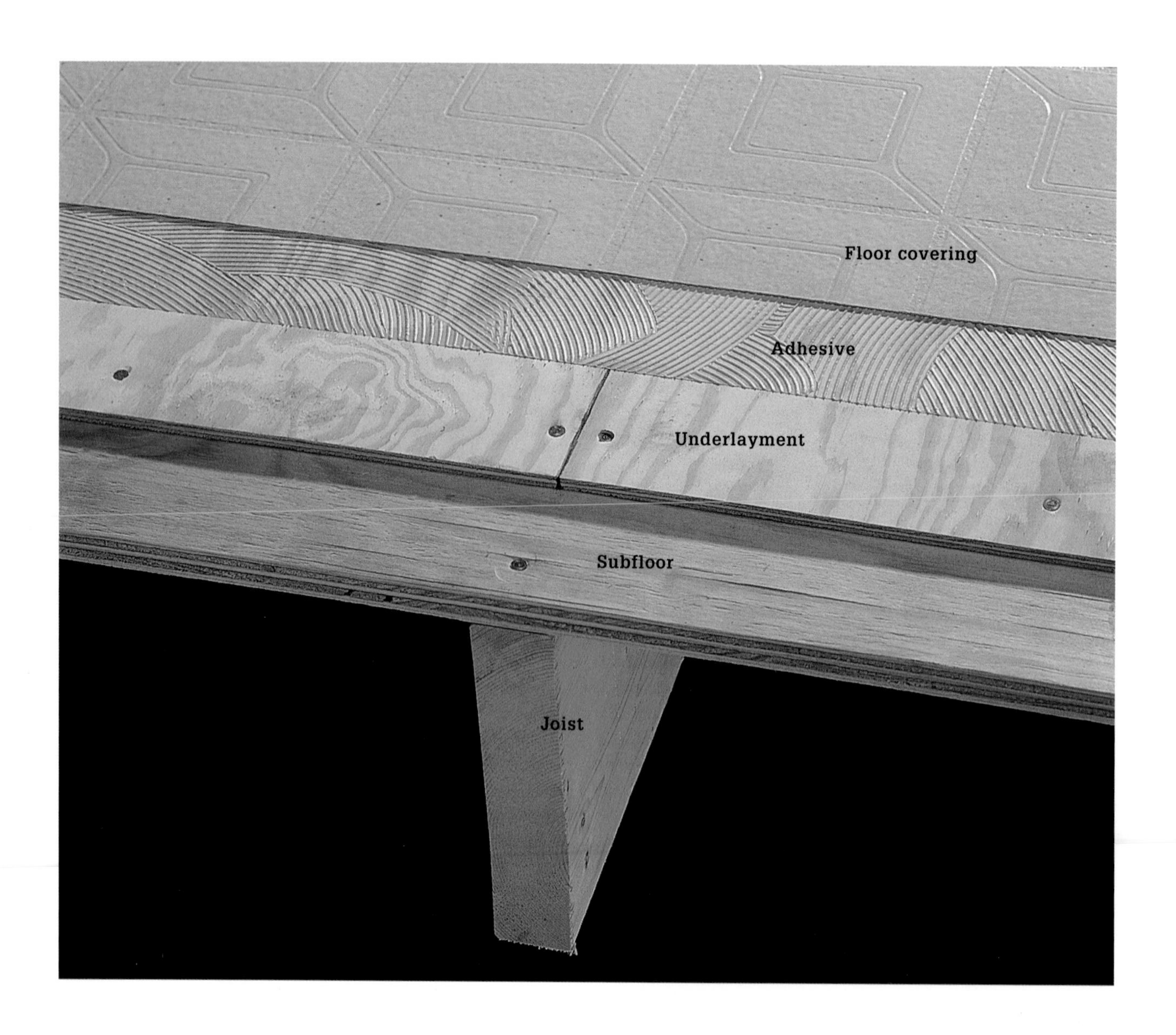

Evaluating an Existing Floor

The first step in preparing for a new floor covering is evaluating your old floor. A careful examination can help you decide whether to repair damaged areas, or replace the flooring altogether.

Evaluating your floor is a three-step process. Begin by identifying the existing floor material and the installation method used. Is your sheet vinyl attached using the full-spread method or the perimeter-bond method? Is your carpet glued down or stretched? Next, check the condition of the floor. Is it securely attached or is it loose in spots? Is it chipped or cracked? Finally, note the height of the existing floor in relation to adjoining floor surfaces. Is it significantly higher than surrounding floors?

A new floor covering or underlayment can often be installed on top of existing flooring. If the existing flooring is not sound or smooth, however, you will have to do some preparation work. Applying a floor leveler (see page 53) is one way to make your existing floor easier to use. More complex preparations may involve removing and replacing the underlayment (see pages 50 to 51) or making spot repairs to the subfloor (see pages 52 to 53).

Warning ▸

Resilient flooring manufactured before 1986 may contain asbestos, which can cause severe lung problems if inhaled. The recommended method for dealing with asbestos-laden flooring is to cover it with an underlayment. If the flooring must be removed, do not do the work yourself. Instead, consult a certified asbestos-abatement contractor.

Determining the number and type of coverings already on your floor is an important early evaluation step. Too many layers of flooring and underlayment can stress floor joists and ultimately cause a new floor to fail. An easy way to check for old flooring is to remove floor vents.

A Quick Guide for Evaluating Your Existing Floor ▸

RESILIENT (VINYL) FLOORING

Option 1: Your existing resilient floor can serve as the foundation for most new floor coverings, including resilient flooring, hardwood, and carpet, but only if the existing surface is relatively smooth and sound. Inspect the existing flooring for loose seams, tears, chips, air bubbles, and other areas where the bond has failed. If these loose spots constitute less than 30% of the total area, you can remove the flooring at these spots and fill the voids with floor-leveling compound. Then, apply embossing leveler to the entire floor and let it dry before laying new resilient flooring.

Option 2: If the original resilient flooring is suspect, you can install new underlayment over the old surface after repairing obviously loose areas.

Option 3: If you're installing ceramic tile, or if the existing surface is in very poor condition, the old resilient flooring should be removed entirely before you install new flooring. If the old flooring was glued down with full-bond adhesive, it's usually easiest to remove both the flooring and underlayment at the same time. If the old underlayment is removed, you must install new underlayment before laying the new flooring.

CERAMIC TILE

Option 1: If the existing ceramic tile surface is relatively solid, new flooring usually can be laid directly over the tile. Inspect tiles and joints for cracks and loose pieces. Remove loose material and fill these areas with a floor-leveling compound. If you're installing resilient flooring, apply an embossing leveler product over the ceramic tile before laying the new flooring. If you're laying new ceramic tile over the old surface, use an epoxy-based thin-set mortar for better adhesion.

Option 2: If more than 10% of the tiles are loose, remove all of the old flooring before installing the new surface. If the tiles don't easily separate from the underlayment, it's best to remove the tile and the underlayment at the same time, then install new underlayment.

HARDWOOD FLOORING

Option 1: If you're installing carpet, you can usually lay it directly over an existing hardwood floor, provided it's a nailed or glued-down surface. Inspect the flooring and secure any loose areas to the subfloor with spiral-shanked flooring nails, then remove any rotted wood and fill the voids with floor-leveling compound before installing the carpet.

Option 2: If you're installing resilient flooring or ceramic tile over nailed hardwood planks or glued-down wood flooring, you can attach new underlayment over the existing hardwood before installing the new flooring.

Option 3: If the existing floor is a "floating" wood or laminate surface with a foam-pad underlayment, remove it completely before laying any type of new flooring.

UNDERLAYMENT & SUBFLOOR

Underlayment must be smooth, solid, and level to ensure a long-lasting flooring installation. If the existing underlayment does not meet these standards, remove it and install new underlayment before you lay new flooring.

Before installing new underlayment, inspect the subfloor for chips, open knots, dips, and loose boards. Screw down loose areas, and fill cracks and dips with floor-leveling compound. Remove and replace any water-damaged areas.

CARPET

Without exception, carpet must be removed before you install any new flooring. For traditional carpet, simply cut the carpet into pieces, then remove the padding and the tackless strips. Remove glued-down cushion-back carpet with a floor scraper, using the same techniques as for removing full-bond resilient sheet flooring (see page 49).

Tips for Evaluating Floors

When installing new flooring over old, measure vertical spaces to make sure enclosed or under-counter appliances will fit once the new underlayment and flooring are installed. Use samples of the new underlayment and floor covering as spacers when measuring.

High thresholds often indicate that several layers of flooring have already been installed on top of one another. If you have several layers, it's best to remove them before installing the new floor covering.

Buckling in solid hardwood floors indicates that the boards have loosened from the subfloor. Do not remove hardwood floors. Instead, refasten loose boards by drilling pilot holes and inserting flooring nails or screws. New carpet can be installed right over a well-fastened hardwood floor. New ceramic tile or resilient flooring should be installed over underlayment placed on the hardwood flooring.

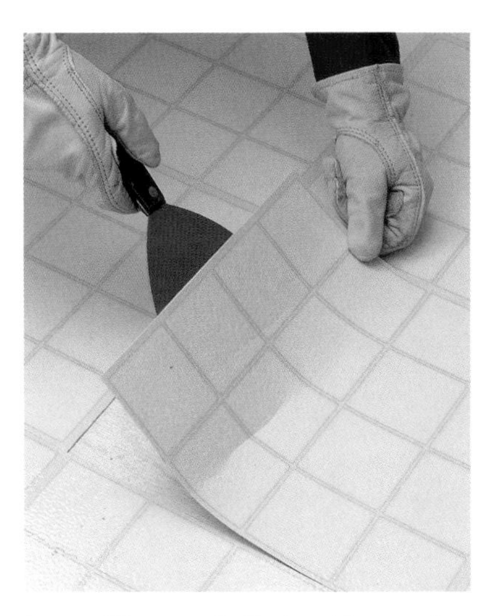

Loose tiles may indicate widespread failure of the adhesive. Use a wallboard knife to test tiles. If tiles can be pried up easily in many different areas of the room, plan to remove all of the flooring.

Air bubbles trapped under resilient sheet flooring indicate that the adhesive has failed. The old flooring must be removed before the new covering can be installed.

Cracks in grout joints around ceramic tile are a sign that movement of the floor covering has caused, or has been caused by, deterioration of the adhesive layer. If more than 10% of the tiles are loose, remove the old flooring. Evaluate the condition of the underlayment (see opposite page) to determine if it also must be removed.

Choosing a Floor Covering

WOOD FLOORS

Wood floors are hard-wearing and durable, yet still look warm and elegant. They hold up well in high-traffic areas and are popular in dining rooms, living rooms, and entryways.

Traditional solid wood planks are the most common type of wood flooring, but there is a growing selection of plywood backed and synthetic-laminate products (also called laminated wood) that are well suited for do-it-yourself installation. Oak and maple are the most common wood species available, and size options include narrow strips, wide planks, and parquet squares. Most wood flooring has tongue-and-groove construction, which helps to provide a strong, flat surface.

In general, hardwood flooring is slightly less expensive than ceramic tile, and laminated products are typically less expensive than solid hardwood. Most types of wood flooring can be installed directly over a subfloor, and sometimes over vinyl flooring. Installation of laminated wood flooring is simple. It can be glued or nailed down, or "floated" on a foam cushion. Parquet squares are typically glued down. Solid hardwood planks must be nailed to the underlayment.

LAMINATE FLOORING

Flooring made from laminated products offers some of the properties of hardwood, such as smooth surfaces with warm colors, but with a very different kind of installation. Instead of being nailed to the underlayment and subfloor, most laminate flooring is fastened edge to edge. It does not require nailing or gluing to the surface below.

Many laminate products are durable enough to carry warranties of 10 years or more, but hardwood remains stronger than most laminates. Better laminate products will have thicker cores, and will feel more like hardwood under your feet. Some feature waterproof joints to reduce the chances of swelling at the core; this becomes important in kitchens and bathrooms.

Since its introduction in the early 1980s, laminate flooring has been a popular choice for homeowners who want the appearance of hardwoods, a choice of finishes, and a material they can install by themselves at a reasonable cost.

VINYL FLOORING

Vinyl flooring, also known as resilient flooring, is a versatile, flexible surface. It is most often found in kitchens and bathrooms, although it can be used almost anywhere. Vinyl flooring is available in both sheets and tiles, in thicknesses ranging from 1/16" to 1/8". Sheets are sold in 6-ft.-wide or 12-ft.-wide rolls, with either a felt or a polyvinyl chloride (PVC) backing, depending on the type of installation. Tiles typically come in 12" squares and are available with or without self-adhesive backing.

Installation is easy. Sheet vinyl with felt backing is glued to the floor using the full-spread method, meaning the entire project area is covered with adhesive. PVC-backed sheet vinyl is glued only along the edges (perimeter-bond method). Tiles are the easiest to install, but because tile floors have a lot of seams, they are less suitable for high-moisture areas. All vinyl flooring must be installed over a smooth underlayment.

Sheet vinyl is priced by the square yard, while tile is priced by the square foot. Cost for either style is comparable to carpet, and less expensive than ceramic tile or hardwood. Prices vary based on the percentage of vinyl in the material, the thickness of the product, and the complexity of the pattern.

CERAMIC TILE

Ceramic tile is a hard, durable, versatile material that is available in a wide variety of sizes, patterns, shapes, and colors. This all-purpose flooring is an excellent choice for areas with high traffic, high moisture, or both. Ceramic tile is commonly used in bathrooms, entryways, and kitchens.

Common ceramic tiles include unglazed quarry tile, glazed ceramic tile, and porcelain mosaic tile. Glass mosaic tiles are available as a colorful alternative to ceramic tiles. In addition, natural stone tiles are sold in several materials, such as marble, slate, and granite. Most floor tiles range from 3⁄16" to 3⁄4" thick.

In general, ceramic tile is more costly than other types of floor coverings, with natural stone tile ranking as the most expensive. Tile is also more time-consuming to install than other materials. However, tile offers the greatest flexibility in design.

Preparation is critical to the success of a tile installation. In high moisture areas, such as bathrooms, tile should be laid over a cementboard underlayment that is fastened to the subfloor. Floors that support tile must be stiff and flat to prevent cracking in the tile surface or grout. Tile is installed following a grid-pattern layout and adhered to the floor with thinset mortar. Gaps between individual tiles are filled with grout, which should be sealed periodically to prevent staining.

CARPET ROLLS

Carpet is a soft, flexible floor covering that is chosen primarily for comfort rather than durability. It is a popular choice for bedrooms, family rooms, and hallways.

Made of synthetic or natural fibers bonded to a mesh backing, carpet is usually sold in 12-ft.-wide rolls. Some types have a cushioned backing, ready for glue-down installation without pads or strips.

The two basic types of carpeting are loop-pile, which is made with uncut loops of yarn to create texture, and cut-pile, which has trimmed fibers for a more uniform appearance. Some carpets contain both types. Carpet is similar in price to vinyl flooring, but costs vary depending on density and fiber. Wool is typically more expensive than synthetics.

Installing carpet is not difficult, but it does involve some special tools and techniques. Tackless strips and padding are installed first; then the carpeting is cut and seamed; finally, it is stretched and secured to tackless strips.

CARPET SQUARES

In the past few years, a new kind of carpet has become very popular: modular squares. These products have most of the qualities people like about roll carpeting—a soft surface and a wide range of colors and patterns. It also offers three things traditional roll carpeting does not: easy installation, simple replacement, and eco-friendliness. The best-known carpet squares are made of recycled material.

Carpet squares need the same kind of underlayment as roll carpeting; it has to be clean and smooth. Unlike carpet rolls, however, the squares do not require stretching or large amounts of adhesive. They are held in place with small adhesive patches. Both the squares and the adhesive patches can be removed and replaced at any time.

Design

No matter what type of floor you install, choose your colors, patterns, and textures carefully. You may have to live with your choices for years—or even decades. Flooring is one of the most visible elements of interior design, which makes it one of the most important.

Flooring can serve as a bold, eye-catching design statement, or it can be an understated background. Whatever approach you take, always consider the design of the adjoining rooms when choosing flooring. Because floors flow from one room to the next, floor coverings offer a convenient medium for creating continuity throughout your home. This does not mean you should use the same floor covering in every room. Simply repeating a color, pattern, or texture can be enough to provide continuity.

The examples on these pages help illustrate how your choice of color, pattern, and texture can affect the look and feel of a room.

Design continuity can be provided by using the same flooring in adjacent rooms. Borders can help define areas, as seen here in the dining area.

Color of flooring influences the visual impact of a room. Bold, bright colors draw attention, while muted colors create a neutral background that doesn't compete for attention. Colors also affect the perceived size of a room. Dark colors are formal in tone, but can make a room look smaller. Light colors are more contemporary, and can make the room seem larger.

Pattern of flooring affects the feeling and tone of a room. In general, subtle patterns lend a more relaxed feel to a room and can make it appear larger. Bold, recurring patterns create excitement and focal points for a room. A flooring pattern must be chosen carefully to ensure it doesn't clash with other patterns in the room.

Texture of flooring contributes to the style of a room. More rugged surfaces, such as slate or Berber carpet, give a room a warm, earthy tone. Smooth and glossy surfaces, such as polished marble tile or hardwood flooring, impart an airy sense of elegance.

Inspiration

The perfect floor is an integral component of any interior design. A well-chosen floor covering will interact with other design elements in the room. The flooring should also be practical and fit the needs of each room. For example, in a kitchen where spills are common, sheet vinyl or ceramic tile is a more practical choice than expensive, deep-pile wool carpet. In a formal dining room, wood parquet is more fitting than resilient tile.

The photos in this section highlight a wide range of flooring types and materials for any room in the home. The following pages are sure to give you new ideas for ways to meet your flooring needs.

Natural hardwoods create an environment of warmth and comfort. From the floor to window sashes and fireplace trim, wood materials combine practicality and luxury.

The medallion in this foyer is sure to catch the eyes of all those who enter this home because of its color and pattern. By incorporating a medallion that is the same material as the surrounding floor and steps, it stands as a subtle and sophisticated detail.

A floor can be the leading design feature in a room from which all other elements in the room follow. In this room the tile floor is the focal point. Smaller tiles around the fireplace echo the pattern set down on the floor surface below.

Refinement and elegance characterize this lightly appointed bathroom, and the ceramic tile underneath it all establishes the tone. By continuing a theme of warm color, plentiful light, and smooth textures from the floor on up, the homeowner has transformed this bathroom into an exclusive spa retreat.

Sometimes, the beauty of simplicity provides the most satisfying results. The ceramic tiles that provide a design base for this entryway will last for years, even decades. Thanks to their neutral color and pattern, they can complement many styles as the homeowners decide to change other elements in the room.

Hardwood strips and planks are the most common wood floor coverings. Hardwood doesn't compete with ornate elements in a room, yet it makes a definitive statement. This floor uses two different woods to create an appealing border around the kitchen island.

This resilient floor combines several subdued colors for a contemporary look. Because the tones are slightly darker than the walls and furniture, the floor grounds the room and serves as a base focal point.

Square and rectangular ceramic tiles can be combined for a sleek, sophisticated look. Laid on a diagonal, the tiles widen a long, narrow room.

Wall-to-wall carpet is very versatile and easily adopts the style and feel of the other room elements. The carpet in this family room takes on a sleek yet comfortable look. By keeping the color slightly lighter than the furniture it acts as a subtle background for the homeowners collections. The monochromatic theme from walls to furniture to carpet enhances the spaciousness of the otherwise small room.

Gentle variations in color—such as wood products or laminates with wood-grain prints—soften the design of a room and create an overall warm tone. Whether you call it blonde or honey or natural, this floor deserves one description: beautiful.

Exclamations of color separated by bold black borders make a dramatic resilient kitchen floor. The visual dividing lines do not correspond exactly with the room transitions, helping bring the spaces together. Look beyond the edges of a room to see more design opportunities.

Iridescent glass tiles, made from recycled material, appear at first glance to have two shades of blue. A closer look reveals that they reflect countless hues, depending how the light strikes them. This works best in small rooms, where the scale of the tiles matches the space.

More than two colors, alternating in an irregular pattern, give this room a casual feel. The classic checkerboard design is muted to create a light and airy space. This attractive, simple design can be made with any material that is sold in squares.

A single color washed over the walls, sinks, and floor generates a cool, smooth surface. The sea blue in this room evokes the calm serenity of water. A soft brown rug provides a sandy beach for the eyes and feet.

Textured concrete floors can be both durable and beautiful. To create a textured surface, you can use stamping forms while the concrete is still wet. To add color as well, apply powdered dyes. This creates the look of stone, ceramic, or even marble. These treatments use a humble material to create a showpiece.

Carpet with just a little detail demands attention. Even though the surface is soft and relatively smooth, that small amount of texture enhances the visual interest of the carpet. In this room it also provides a gentle contrast to the plush bedding and sheer window coverings.

Hexagonal tiles have been available for centuries, and still provide an eye-catching effect. These tiles also have broad variations in color, giving them added visual appeal. Installation and maintenance are the same as for square ceramic tiles. Remember, not every tile floor needs to be made of squares or rectangles.

A smooth tile floor like this has variations in color and rough-cut edges that give it character. The mottled rust and verdigris are reminiscent of aging bronze sculptures, while the grout lines look like the work of ancient master masons. In real life, however, you could install this floor yourself—today.

A checkerboard inside a checkerboard creates contrast in color, size, and shape. The effect draws your eye across the surface, while subdued brown tones provide an understated foundation for the surprise of bold red walls.

A monotone scheme is perfect for this small space. Consistent color on the walls and floor makes the room feel larger, but the random visual pattern keeps it from feeling dull. The tumbled stones create a natural texture underfoot. By taking a little extra care and expense, the homeowner has created a room that makes part of every day life feel like a vacation.

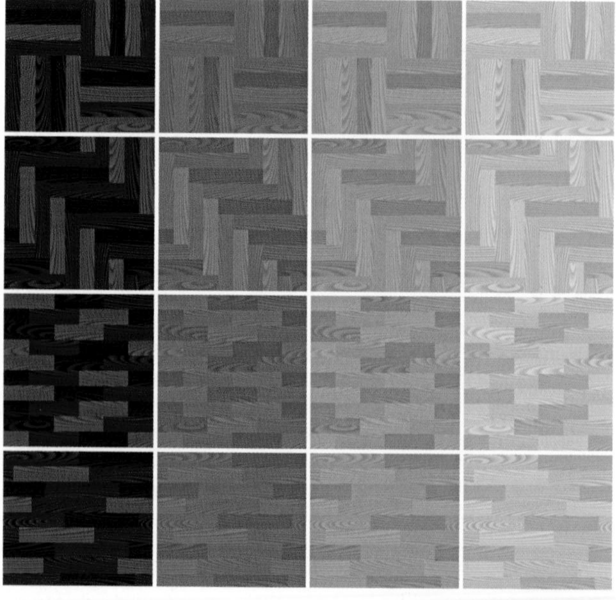

Bricklaying patterns borrowed from masonry can work for other materials, too. A running-bond pattern has parallel courses overlapping at the halfway point of each rectangle, while a herringbone pattern places them in rows of nestled V-shapes. From top row to bottom row: basket weave, herring bone, running-bond, stretcher-bond.

The stripes in this carpet run the same direction, but they vary in width and color, making the design less predictable. By using a carpet with several colors that work well together, the homeowner can change the room furnishings without having to switch the floor covering as well. A bold carpet design like this is best complemented by solid colors on the walls and furniture.

Regular and random at the same time, this floor makes a compromise by marrying solid and mottled blocks. Every row has the same width, but individual tiles within each row are not as structured. The result is a design that both follows a rhythm and breaks it.

A smooth blue floor is just about the coolest foundation possible for a bedroom. While leaving the rest of the white room airy and light, the checkerboard and solid blue field create a sense of quiet calm.

The warm (tans) and cool (whites) color combinations of this bathroom balance cleanliness and comfort. Despite using only right angles on the fixtures and walls, this sparsely appointed room maintains an elegant, informal appearance.

Limestone and glass establish a calm yet sophisticated atmosphere in this bathroom. Subtle variations in tile color provide visual interest; colored grout keeps the color family together.

A deep, rich stain on this hardwood floor gives the room a solid color foundation for the warm autumn colors on the walls and ceiling. Incandescent lights enhance the warm glow in the room.

Hard angles and surfaces in this room reflect sound sharply, but the soft greens make a peaceful setting. While conventional wisdom holds that small tiles only work in small spaces, this stunning bathroom proves an exception.

Mixed materials can bring the best of both sound-reflecting and sound-absorbing qualities. In this example, a highly reflective hardwood surface becomes hushed by a small area rug. The hardwood's beauty remains visible, and the area rug complements the décor.

Soft carpet has sound-absorbing properties, which are perfectly suited to this den of solitude. Unwanted noise disappears quickly, and only the sounds you choose remain. Even the furnishings have soft edges, and the colors are muted.

Concrete floors are now being welcomed into rooms beyond just their traditional use in basements and garages. A variety of surface treatments can transform this utilitarian material into a fashionable floor. One characteristic of concrete won't change, though: It's still hard, and it reflects sound. Offset these drawbacks with soft furniture and window coverings.

A luxurious carpet becomes a practical indulgence when the only shoes in the room are your bedroom slippers. Of course, you will want it to be durable, even if you treat it gently. Wool fibers are among the most long-wearing versions available, and Berber loops are especially comfortable on bare feet.

Ceramic tile holds up well under heavy use. But it can be tiring to stand on for long periods, and may feel cold underfoot. New layered products, combining ceramic and resilient materials, give you the look and feel of ceramic tile, with the warmth and easy installation of vinyl tile.

Quarry tile has been a favorite of commercial kitchens for decades. It stands up to heavy use day and night, and is relatively easy to clean. It also has a naturally warm appearance that gives any kitchen a soft glow. Other popular places for quarry tile are bathrooms and spas.

Ceramic tile is the preferred flooring material for bathrooms. It is easy to clean and dry, and it withstands frequent exposure to water. Manufacturers often create matched families of tile designs for floors, walls, and countertops.

A temporary covering can help preserve the permanent surface underneath when an area of flooring gets rough treatment from the elements, foot traffic, or both. This whimsical rug adds a dash of bright color and playful humor to a dining room while reducing wear on the floor below.

Linoleum floors with custom designs, like this one, are especially sensitive to uneven surfaces or debris underneath. Even after it's in place, linoleum is not trouble-free. It can be cleaned with a variety of floor care products, but never with anything containing ammonia.

Leather may be the most fragile of all flooring materials. It's luxurious to a fault, but vulnerable to abrasion, tears, cuts, and damage from any sharp or rough object. The tile floor shown here has a faux leather appearance, but maintains the durability of ceramic tile.

Hardwood floors can last for decades with regular care. If a hardwood floor is used heavily, it may need repairing. Luckily, most hardwoods can be sanded down and refinished at least two or three times before they must be replaced.

This luxury vinyl tile offers the rugged beauty of cut stone, but with easy care. Its solid vinyl construction is more durable than traditional composition tile, and its top wear layer resists abrasion. Its greatest advantage over stone and ceramic tile is that vinyl never needs mortar or grout.

Bamboo floors have become very popular in the past few years. It's a grass, not a wood, so it grows quickly. And the process of turning bamboo into flooring is ecologically friendly. Most important, it has the best qualities of hardwood with greater durability, value, and ease of installation.

Cork flooring can be durable enough for use in high-traffic areas such as kitchens. Its slightly soft feel underfoot makes it an ideal surface for long periods of standing. And, if you drop a glass on it, both the glass and the floor may remain intact.

Palm is another fast-growing plant. Like bamboo, it lacks the strength to be used like a traditional hardwood. However, it can be processed the same way as bamboo to produce an equally hard-wearing floor covering. This plank flooring has all the warmth of hardwood, but the material is renewable.

Linoleum, a product of linseed oil, is one of the oldest renewable flooring materials to be mass-produced. It has enjoyed a revival of popularity in recent years, thanks to its eco-friendly properties and design flexibility. Linoleum is often difficult to handle, but the finished results are visually striking.

Carpet squares made from recycled waste soften hard surfaces, add contrasting color to plain rooms, and offer countless design possibilities, all without consuming new raw materials. Depending on your design, you can place them from one edge of a room to the other, or set them in the middle as a rug,

Scrap rubber never biodegrades, which makes it a persistent headache for municipal waste departments. But that same quality makes reclaimed rubber an excellent flooring material for high-traffic areas. These tiles are made from rubber products that would otherwise be landfill. They go much better in a garage, basement, or breezeway.

Hardwoods make some of the best flooring materials—sometimes more than once. This rich, dark maple was reclaimed during the demolition of an old property. It may not be a common item in large home improvement stores, but you can find reclaimed hardwoods through some flooring stores and specialty lumberyards.

Recycled materials give old materials—and your home—new life. Here, from left to right, tile made of recycled glass, aluminum, and brass are shown.

Project Preparation

Before your new floor goes in, your old floor will probably need to be taken out and the subfloor will need to be carefully prepared for a finished surface. Project preparation is just as important as installing your floor covering and requires the same attention to detail.

Removing old floors, installing new subfloors or underlayments, and filling in cracks and joints isn't the most glamorous job in the world, but it's an investment that will reap big rewards when your flooring project is complete.

If your new floor is part of a larger home improvement project, removing the existing floor is one of the first steps in the overall project, while installing the new floor is one of the last steps in the process. All other demolition and construction should be finished in the room before the floor is installed to avoid damaging the surface.

In this Chapter:

Preparation Tools & Materials

Tools for flooring removal and surface preparation include: a power sander (A), jamb saw (B), putty knife (C), floor roller (D), circular saw (E), hammer (F), hand maul (G), reciprocating saw (H), cordless drill (I), flat-edged trowel (J), notched trowel (K), stapler (L), cat's paw (M), flat pry bar (N), heat gun (O), masonry chisel (P), crowbar (Q), nippers (R), wallboard knife (S), wood chisel (T), long-handled floor scraper (U), Phillips screwdriver (V), standard screwdriver (W), utility knife (X), and carpenter's level (Y).

Measuring the Room

Before ordering your floor covering, determine the total square footage of your room. To do this, divide the room into a series of squares and rectangles that you can easily measure. Be sure to include all areas that will be covered, such as closets and space under your refrigerator and other movable appliances.

Measure the length and width of each area in inches, then multiply the length times the width. Divide that number by 144 to determine your square footage. Add all of the areas together to figure the square footage for the entire room, then subtract the areas that will not be covered, such as cabinets and other permanent fixtures.

When ordering your floor covering, be sure to purchase 10 to 15% extra to allow for waste and cutting. For patterned flooring, you may need as much as 20% extra.

Measure the area of the project room to calculate how much flooring you will need.

How to Measure Your Room

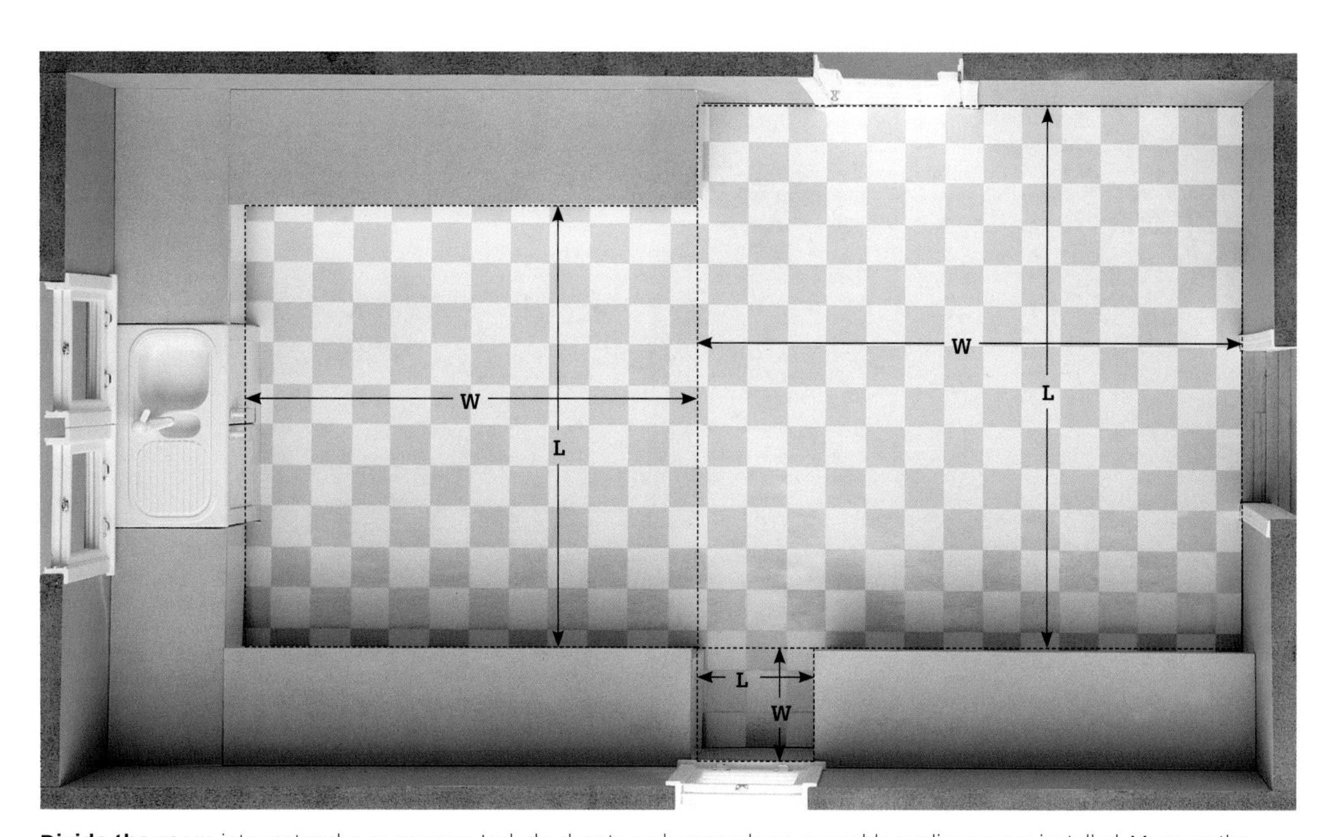

Divide the room into rectangles or squares. Include closets and areas where moveable appliances are installed. Measure the length and width of each area in inches, then multiply the length times the width. Divide that number by 144 to determine your square footage.

Removing Floor Coverings

When old floor coverings must be removed, as is the case with many projects, thorough and careful removal work is essential to the quality of the new flooring installation.

The difficulty of flooring removal depends on the type of floor covering and the method that was used to install it. Carpet and perimeter-bond vinyl are generally very easy to remove, and removing vinyl tiles is also relatively simple. Full-spread sheet vinyl can be difficult to remove, however, and removing ceramic tile entails a lot of work.

With any removal project, be sure to keep your tool blades sharp, and avoid damaging the underlayment if you plan to reuse it. If you'll be replacing the underlayment, it may be easier to remove the old underlayment along with the floor covering.

Use a floor scraper to remove resilient floor coverings and to scrape off leftover adhesives or backings. The long handle provides leverage and force, and it allows you to work in a comfortable standing position. A scraper will remove most of the flooring, but you may need other tools to finish the job.

Tools & Materials ▸

- Floor scraper
- Utility knife
- Spray bottle
- Wallboard knife
- Wet/dry vacuum
- Heat gun
- Hand maul
- Masonry chisel
- Flat pry bar
- End-cutting nippers
- Liquid dishwashing detergent
- Scrap wood

OPTIONS FOR REMOVING OLD FLOORING

Remove the floor covering only. If the underlayment is sturdy and in good condition, you can usually get by with simply scraping off the floor covering, then cleaning and reusing the existing underlayment.

Remove the floor covering and underlayment. If the underlayment is questionable or substandard, or if the floor covering is bonded to the underlayment, remove the flooring and underlayment together. Taking up both layers at once saves time.

How to Prepare the Installation Space

Disconnect and remove all appliances. When bringing the appliances back into the room, protect your new floor by placing cardboard or a heavy cloth on the floor and in front of the appliance locations. Before setting the appliances in place, make sure the floor adhesives are properly cured.

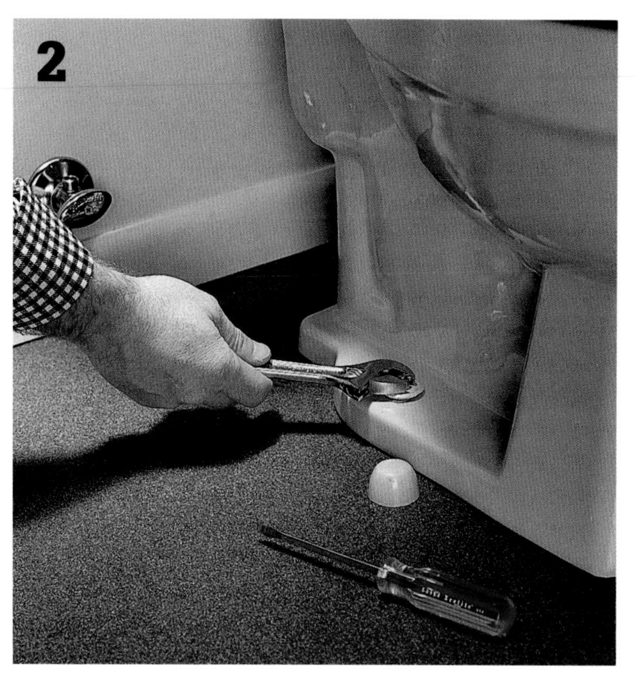

Remove the toilet and other floor-mounted fixtures before installing a bathroom floor. Turn off and disconnect the water supply line, then remove the bolts holding the toilet on the floor.

Ventilate the project room, especially when applying adhesives or removing old flooring. Placing a box fan in an open window will help draw dust and noxious fumes from the work area.

Cover entryways with sheet plastic to contain dust and debris while you remove the old floor.

(continued)

Cover heat and air vents with sheet plastic and masking tape to prevent dust and debris from entering ductwork.

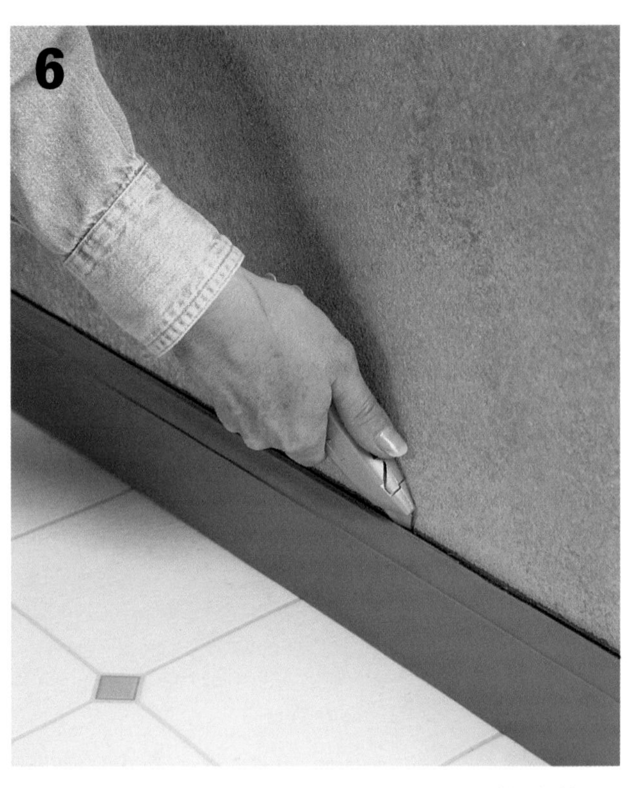

Cut the paint away from the baseboard with a utility knife.

Place a scrap board against the wall to avoid damaging the drywall. Remove the baseboard using a pry bar placed against the scrap board. Pry the baseboard at all nail locations. Number the baseboards as they are removed.

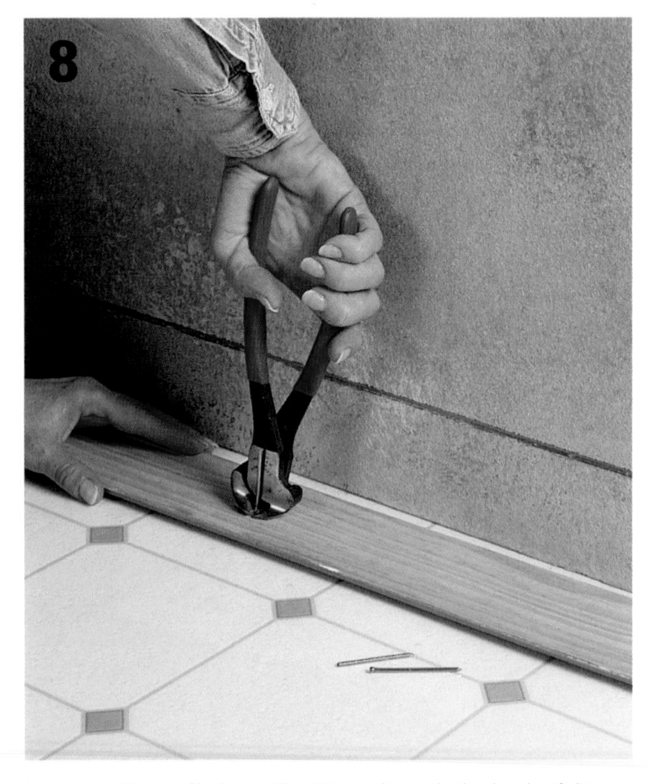

Remove the nails by pulling them through the back of the baseboard with nippers.

How to Remove Sheet Vinyl

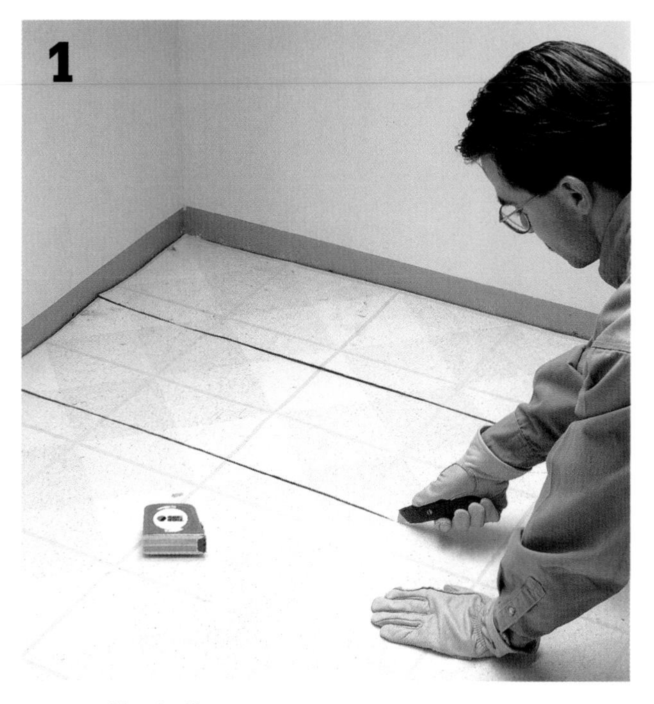

Use a utility knife to cut the old flooring into strips about a foot wide to make removal easier.

Pull up as much flooring as possible by hand. Grip the strips close to the floor to minimize tearing.

Cut stubborn sheet vinyl into strips about 6" wide. Starting at a wall, peel up as much of the floor covering as possible. If the felt backing remains, spray a solution of water and liquid dishwashing detergent under the surface layer to help separate the backing. Use a wallboard knife to scrape up particularly stubborn patches.

Scrape up the remaining sheet vinyl and backing with a floor scraper. If necessary, spray the backing with the soap solution to loosen it. Sweep up the debris, then finish the cleanup using a wet/dry vacuum. *Tip: Fill the vacuum with about an inch of water to help contain dust.*

How to Remove Vinyl Tile

Starting at a loose seam, use a long-handled floor scraper to remove tiles. To remove stubborn tiles, soften the adhesive with a heat gun, then use a wallboard knife to pry up the tile and scrape off the underlying adhesive.

Remove stubborn adhesive or backing by wetting the floor with a mixture of water and liquid dishwashing detergent, then scrape it with a floor scraper.

How to Remove Ceramic Tile

Knock out tile using a hand maul and masonry chisel. If possible, start in a space between tiles where the grout has loosened. Be careful when working around fragile fixtures, such as drain flanges, to prevent damage.

If you plan to reuse the underlayment, use a long-handled floor scraper to remove any remaining adhesive. You may have to use a belt sander with a coarse sanding belt to grind off stubborn adhesive.

How to Remove Carpet

Using a utility knife, cut around metal threshold strips to free the carpet. Remove the threshold strips with a flat pry bar.

Cut the carpet into pieces small enough to be easily removed. Roll up the carpet and remove it from the room, then remove the padding. *NOTE: Padding is often stapled to the floor and usually comes up in pieces as you roll it.*

Using end-cutting nippers or pliers, remove all of the staples from the floor. If you plan to lay new carpet, leave the tackless strips in place unless they are damaged.

Variation: To remove glued-down carpet, cut it into strips with a utility knife, then pull up as much material as you can. Scrape up the remaining cushion material and adhesive with a floor scraper.

Removing Underlayment

Flooring contractors routinely remove the underlayment along with the floor covering before installing new flooring. This saves time and makes it possible to install new underlayment that's ideally suited to the new flooring. Do-it-yourselfers using this technique should make sure to cut the flooring into pieces that can be easily handled.

Warning ▸

This floor removal method releases flooring particles into the air. Be sure the flooring you are removing does not contain asbestos.

Tools & Materials ▸

- Goggles
- Gloves
- Circular saw with carbide-tipped blade
- Flat pry bar
- Reciprocating saw
- Wood chisel
- Hammer
- Protective ear and eye gear
- Dust mask

Removal Tip ▸

Examine fasteners to see how the underlayment is attached. Use a screwdriver to expose the heads of the fasteners. If the underlayment has been screwed down, you'll need to remove the floor covering and then unscrew the underlayment.

Remove the underlayment and floor covering as though they're a single layer. This is an effective removal strategy with any floor covering that's bonded to the underlayment.

How to Remove Underlayment

Adjust the cutting depth of a circular saw to equal the combined thickness of your floor covering and underlayment. Using a carbide-tipped blade, cut the floor covering and underlayment into squares measuring about 3 feet square. Be sure to wear safety goggles and gloves.

Use a reciprocating saw to extend the cuts to the edges of the walls. Hold the blade at a slight angle to the floor and be careful not to damage walls or cabinets. Don't cut deeper than the underlayment. Use a wood chisel to complete cuts near cabinets.

Separate the underlayment from the subfloor using a flat pry bar and hammer. Remove and discard the sections of underlayment and floor covering immediately, watching for exposed nails.

Variation: If your existing floor is ceramic tile over plywood underlayment, use a hand maul and masonry chisel to chip away the tile along the cutting lines before making cuts.

Repairing Subfloors

A solid, securely fastened subfloor minimizes floor movement and squeaks. It also ensures that your new floor covering will last a long time.

After removing the old underlayment, inspect the subfloor for loose seams, moisture damage, cracks, and other flaws. If your subfloor is made of dimension lumber rather than plywood, you can use plywood to patch damaged sections. If the plywood patch doesn't reach the height of the subfloor, use floor leveler to raise its surface to match the surrounding area.

Tools & Materials ▸

- Flat-edged trowel
- Straightedge
- Framing square
- Drill
- Circular saw
- Cat's paw
- Wood chisel
- Hammer
- Tape measure
- 2" deck screws
- Carpenter's leveler
- Plywood
- 2 × 4 lumber
- 10d common nails
- Protective gloves

How to Apply Floor Leveler

Floor leveler is used to fill in dips and low spots in plywood subfloors. Mix the leveler according to directions from the manufacturer, adding a latex or acrylic bonding agent for added strength.

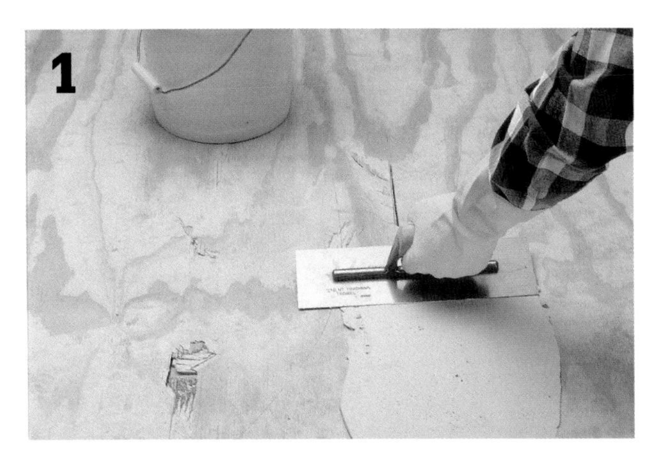

Mix the leveler according to the manufacturer's directions, then spread it onto the subfloor using a finishing trowel. Build up the leveler in thin layers to avoid overfilling the area, allowing each layer to dry before applying the next.

Use a straightedge to make sure the filled area is level with the surrounding area. If necessary, apply more leveler. Allow the leveler to dry, then shave off any ridges with the edge of a trowel, or if necessary, sand it smooth.

How to Replace a Section of Subfloor

Use a framing square to mark a rectangle around the damage, making sure two sides of the rectangle are centered over floor joists. Remove nails along the lines, using a cat's paw. Make the cuts using a circular saw adjusted so the blade cuts through the subfloor only. Use a chisel to complete cuts near walls.

Remove the damaged section, then nail two 2 × 4 blocks between the joists, centered under the cut edges for added support. If possible, end-nail the blocks from below. Otherwise, toe-nail them from above, using 10d nails.

Measure the cut-out section, then cut a patch to fit. Use material that's the same thickness as the original subfloor. Fasten the patch to the joists and blocks using 2" deck screws spaced about 5" apart.

Repairing Joists

A severely arched, bulged, cracked, or sagging floor joist can only get worse over time, eventually deforming the floor above it. Correcting a problem joist is an easy repair and makes a big difference in your finished floor. It's best to identify problem joists and fix them before installing your underlayment and new floor covering.

One way to fix joist problems is to fasten a few new joists next to a damaged floor joist in a process called sistering. When installing a new joist, you may need to notch the bottom edge so it can fit over the foundation or beam. If that's the case with your joists, cut the notches in the ends no deeper than ⅛" of the actual depth of the joist.

Tools & Materials ▸

- 4-ft. level
- Reciprocating saw
- Hammer
- Chisel
- Adjustable wrench
- Tape measure
- Ratchet wrench
- 3" lag screws with washers
- Framing lumber
- 16d common nails
- Hardwood shims
- Metal jack posts

How to Repair a Bulging Joist

Find the high point of the bulge in the floor using a level. Mark the high point and measure the distance to a reference point that extends through the floor, such as an exterior wall or heating duct.

Use the measurement and reference point from the last step to mark the high point on the joist from below the floor. From the bottom edge of the joist, make a straight cut into the joist just below the high point mark using a reciprocating saw. Make the cut ¾ of the depth of the joist. Allow several weeks for the joist to straighten.

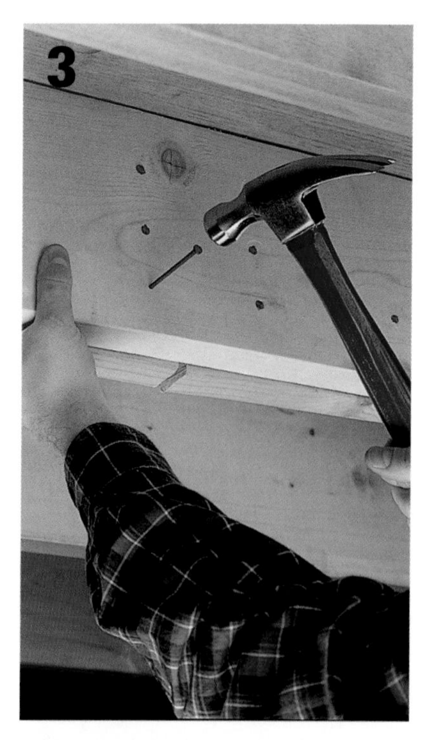

When the joist has settled, reinforce it by centering a board of the same height and at least 6 ft. long next to it. Fasten the board to the joist by driving 12d common nails in staggered pairs about 12" apart. Drive a row of three nails on either side of the cut in the joist.

How to Repair a Cracked or Sagging Joist

Identify the cracked or sagging joist before it causes additional problems. Remove any blocking or bridging above the sill or beam where the sister joist will go.

Place a level on the bottom edge of the joist to determine the amount of sagging that has occurred. Cut a sister joist the same length as the damaged joist. Place it next to the damaged joist with the crown side up. If needed, notch the bottom edge of the sister joint so it fits over the foundation or beam.

Nail two 6-ft. 2 × 4s together to make a cross beam, then place the beam perpendicular to the joists near one end of the joists. Position a jack post under the beam and use a level to make sure it's plumb before raising it.

Raise the jack post by turning the threaded shaft until the cross beam is snug against the joists. Position a second jack post and cross beam at the other end of the joists. Raise the posts until the sister joist is flush with the subfloor. Insert tapered hardwood shims at the ends of the sister joist where it sits on the sill or beam. Tap the shims in place with a hammer and scrap piece of wood until they're snug.

Drill pairs of pilot holes in the sister joist every 12", then insert 3" lag screws with washers in each hole. Cut the blocking or bridging to fit and install it between the joists in its original position.

Preparing the Base

Every floor in your house is made of layers. Under the floor covering, you may have carpet padding, isolation membrane, hardened adhesives, or previous flooring materials. To create a lasting floor, you first have to get to the underlayment and, if necessary, the subfloor beneath it.

Plan the removal in steps. Begin by clearing the room, redirecting traffic, and protecting adjacent spaces from debris by sealing off doors and windows with plastic. Don't forget to remove baseboards and any other trimwork that touches the floor. Then you should be ready to take out the top floor covering.

As you work your way down through the layers, pay attention to their condition. A damaged underlayment is not worth preserving. For that matter, if you're taking up a ceramic tile floor and the work is difficult and tedious, you may decide to take out the underlayment to save time.

As with all demolition projects, be sure to protect your eyes, ears, and lungs. If possible, place a fan in a window to blow dust out of the room. Wear gloves to prevent splinters and blisters. And remember to stop when you're tired. That's when accidents are most likely, and you're the least prepared to handle them. Take your time. This is just the beginning.

In this Chapter

Installing Underlayment

Underlayment is a layer of sheeting screwed or nailed to the subfloor to provide a smooth, stable surface for the floor covering. The type of underlayment you choose depends in part on the type of floor covering you plan to install. Ceramic and natural stone tile floors usually require an underlayment that stands up to moisture, such as cementboard. For vinyl flooring, use a quality-grade plywood; most warranties are void if the flooring is installed over substandard underlayments. If you want to use your old flooring as underlayment, apply an embossing leveler to prepare it for the new installation. Most wood flooring and carpeting do not require underlayment and are often placed directly on a plywood subfloor.

When you install new underlayment, attach it securely to the subfloor in all areas, including under movable appliances. Notch the underlayment to fit the room's contours. Insert the underlayment beneath door casings and moldings. Once the underlayment is installed, use a latex patching compound to fill gaps, holes, and low spots. This compound is also used to cover screw heads, nail heads, and seams in underlayment. Some compounds include dry and wet ingredients that need to be mixed, while others are premixed. The compound is applied with a trowel or wallboard knife.

Tools & Materials ▸

- Drill
- Circular saw
- Wallboard knife
- Power sander
- ¼" notched trowel
- Straightedge
- Utility knife
- Jig saw with carbide-tipped blade
- ⅛" notched trowel
- Flooring roller
- Underlayment
- 1" deck screws
- Floor-patching compound
- Latex additive
- Thin-set mortar
- 1½" galvanized deck screws
- Fiberglass-mesh wallboard tape

How to Install Plywood Underlayment

Plywood is the most common underlayment for vinyl flooring and some ceramic tile installations. For vinyl, use ¼" exterior-grade, AC plywood. This type has one smooth side for a quality surface. Wood-based floor coverings, like parquet, can be installed over lower-quality exterior-grade plywood. For ceramic tile, use ½" AC plywood. When installing plywood, leave ¼" expansion gaps at the walls and between sheets.

Install a full sheet of plywood along the longest wall, making sure the underlayment seams are not aligned with the subfloor seams. Fasten the plywood to the subfloor using 1" deck screws driven every 6" along the edges and at 8" intervals in the field of the sheet.

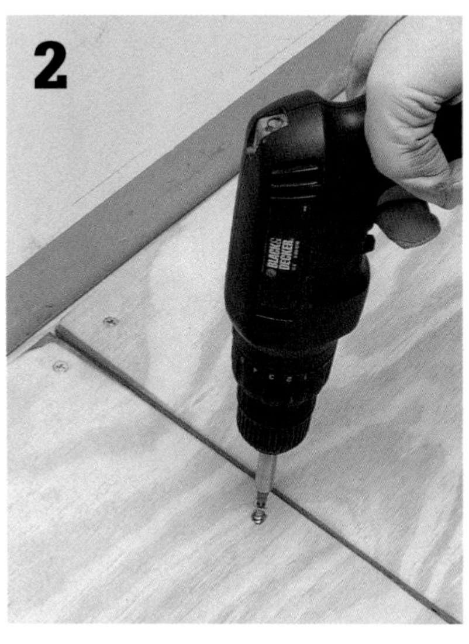

Continue fastening sheets of plywood to the subfloor, driving the screw heads slightly below the underlayment surface. Leave ¼" expansion gaps at the walls and between sheets. Offset seams in subsequent rows.

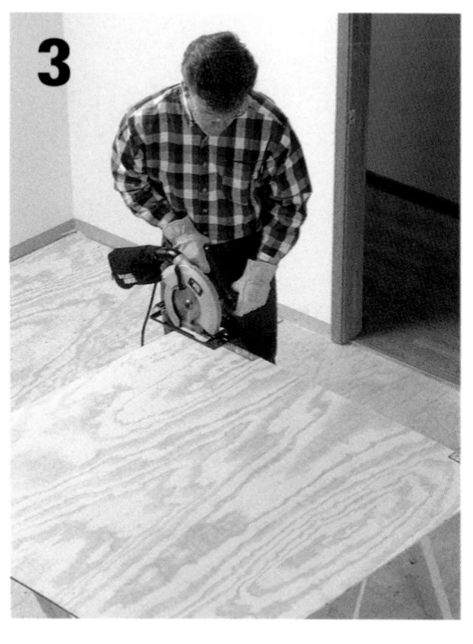

Using a circular saw or jig saw, notch the plywood to meet the existing flooring in doorways. Fasten the notched sheets to the subfloor.

Mix floor-patching compound and latex or acrylic additive following the manufacturer's directions. Spread it over seams and screw heads, using a wallboard knife.

Let the patching compound dry, then sand the patched areas, using a power sander.

How to Install Cementboard

Ceramic and natural stone tile floors usually require an underlayment that stands up to moisture, such as cementboard. Fiber/cementboard is a thin, high-density underlayment used under ceramic tile and vinyl flooring in situations where floor height is a concern. Cementboard is used only for ceramic tile or stone tile installations. It remains stable even when wet, so it is the best underlayment to use in areas that are likely to get wet, such as bathrooms. Cementboard is more expensive than plywood, but a good investment for a large tile installation.

Mix thin-set mortar according to the manufacturer's directions. Starting at the longest wall, spread the mortar on the subfloor in a figure eight pattern using a ¼" notched trowel. Spread only enough mortar for one sheet at a time. Set the cementboard on the mortar with the rough side up, making sure the edges are offset from the subfloor seams.

Fasten the cementboard to the subfloor using 1¼" cementboard screws driven every 6" along the edges and 8" throughout the sheet. Drive the screw heads flush with the surface. Continue spreading mortar and installing sheets along the wall. *OPTION: If installing fiber/cementboard underlayment, use a 3⁄16" notched trowel to spread the mortar, and drill pilot holes for all screws.*

Cut cementboard pieces as necessary, leaving an ⅛" gap at all joints and a ¼" gap along the room perimeter. For straight cuts, use a utility knife to score a line through the fiber-mesh layer just beneath the surface, then snap the board along the scored line.

To cut holes, notches, or irregular shapes, use a jig saw with a carbide-tipped blade. Continue installing cementboard sheets to cover the entire floor.

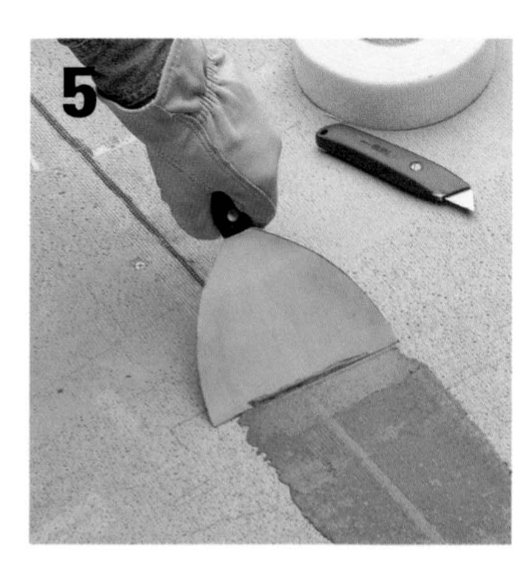

Place fiberglass-mesh wallboard tape over the seams. Use a wallboard knife to apply thin-set mortar to the seams, filling the gaps between sheets and spreading a thin layer of mortar over the tape. Allow the mortar to set for two days before starting the tile installation.

How to Install Isolation Membrane

Isolation membrane is used to protect ceramic tile installations from movement that may occur on cracked concrete floors. This product is used primarily for covering individual cracks, but it can be used over an entire floor. Isolation membrane is also available in a liquid form that can be poured over the project area.

Thoroughly clean the subfloor, then apply thin-set mortar with a ⅛" notched trowel. Start spreading the mortar along a wall in a section as wide as the membrane and 8 to 10 ft. long. *NOTE: For some membranes, you must use a bonding material other than mortar. Read and follow manufacturer's directions.*

Roll out the membrane over the mortar. Cut the membrane to fit tightly against the walls, using a straightedge and utility knife.

Starting in the center of the membrane, use a heavy floor roller to smooth out the surface toward the edges. This frees trapped air and presses out excess bonding material.

Repeat steps 1 through 3, cutting the membrane as necessary at the walls and obstacles, until the floor is completely covered with membrane. Do not overlap the seams, but make sure they're tight. Allow the mortar to set for two days before installing the tile.

How to Install Soundproofing Underlayment

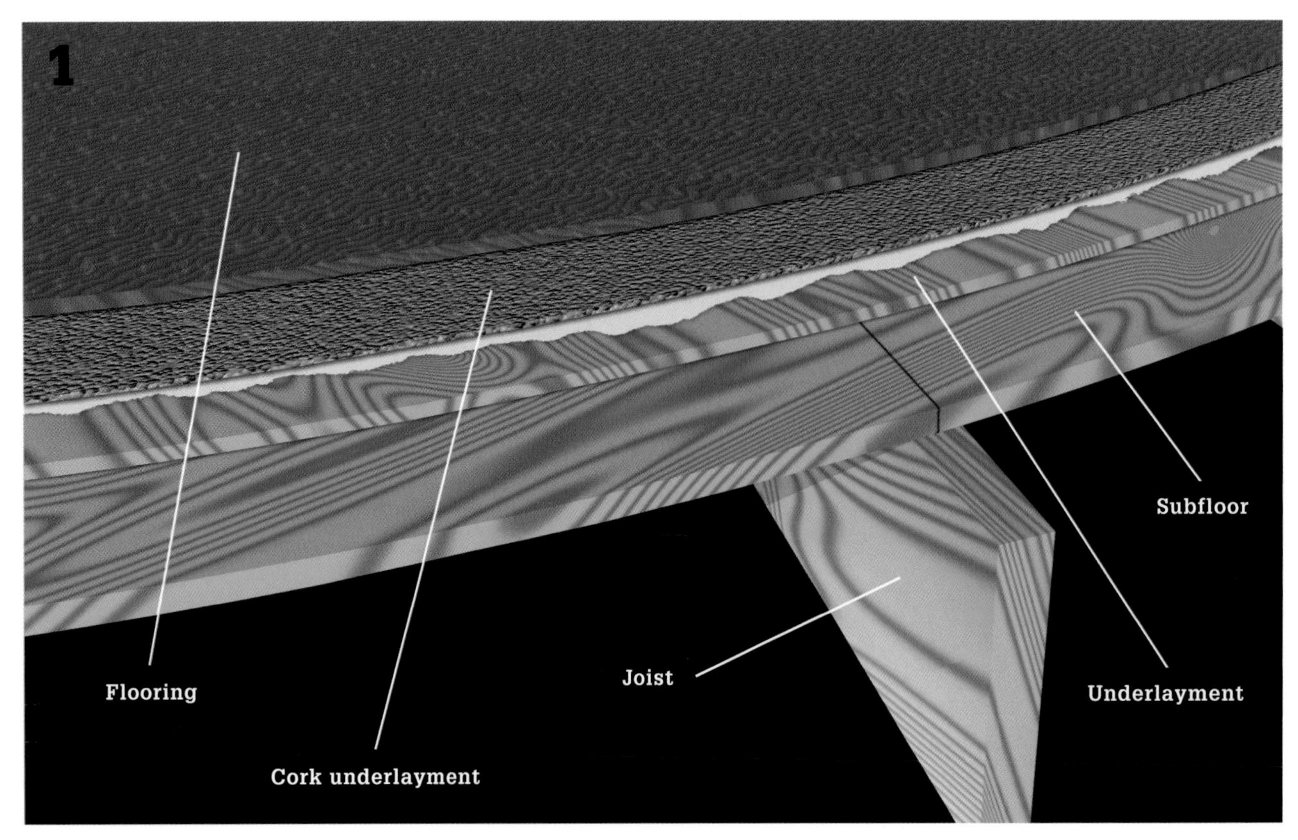

Soundproofing underlayment fits just under the floor covering. Remove floor covering, if necessary (see removal how-to steps starting on page 40).

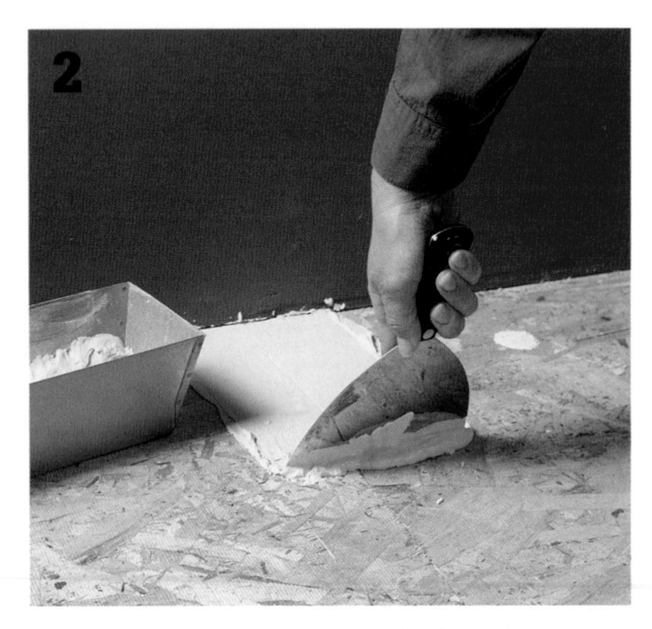

Patch all holes, cracks, and joints in the plywood underlayment with cement-based compound, using a wallboard knife. The patches must be dry and the subfloor clean before continuing.

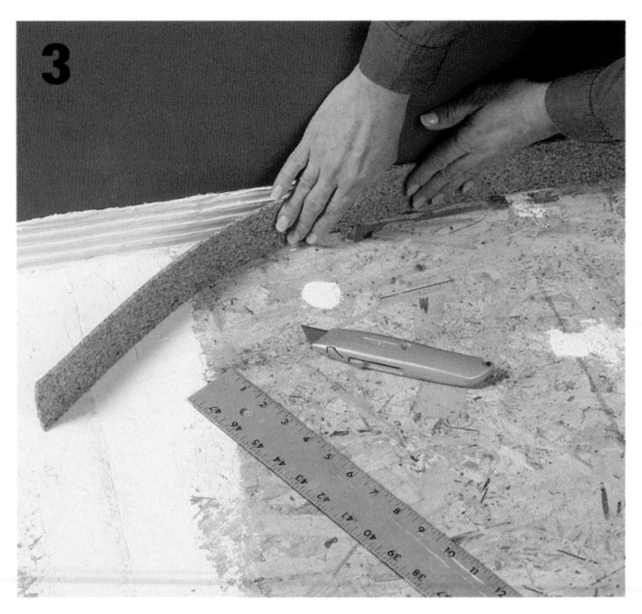

Cut the cork into 2" strips using a straightedge and utility knife. Using a manufacturer approved adhesive, apply the strips to the base of the walls so the bottom edge sits on the floor. Press firmly to eliminate air bubbles.

Unroll the cork the length of the room so the curled side is face down. Butt it against the 2" strips. Be careful not to crease the cork; it's flexible enough to curl, but not to fold.

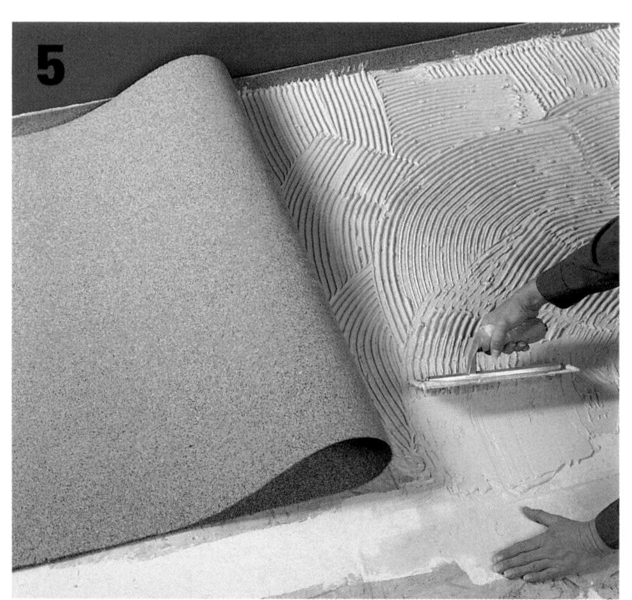

Pull back at least half of the roll. Apply adhesive to the plywood underlayment and spread it out, using a notched trowel. Replace the cork over the adhesive.

Roll the cork front to back and side to side using a floor roller. Repeat these steps to adhere the other half of the cork to the plywood underlayment. Cover the rest of the floor the same way. Butt joints tightly together, but don't overlap them.

Installing Raised Underlayment Panels

Concrete floors are practical and durable—and generally cold and uncomfortable. For a fast and easy makeover, you can now find raised underlayment panels that simply rest on the concrete and provide a surface for other flooring materials. The tongue-and-groove plywood panels have dimpled plastic on the bottom. This allows air to circulate underneath so that the concrete stays dry, and insulates the flooring above. The assembled panels can support laminates, resilient sheets, or tiles. And you can install them in a weekend.

How to Install Raised Underlayment Panels

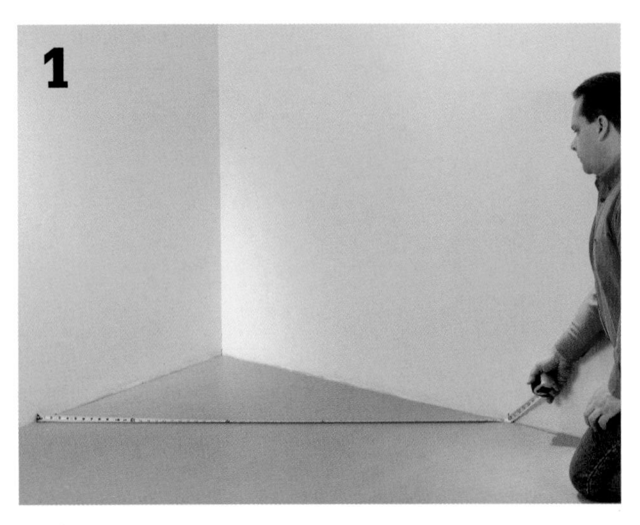

Start at one corner, and measure the length and width of the room from that starting point. Calculate the number of panels you will need to cover the space in both directions. If the starting corner is not square, trim the first row to create a straight starting line.

Create an expansion gap around the edges. Place ¼" spacers at all walls, doors, and other large obstacles. To make your own spacers, cut sheets of ¼" plywood to the thickness of the panels, and hold them in place temporarily with adhesive tape.

Dry-lay a row of panels across the room. If the last row will be less than 6" wide, balance it by trimming the first panel or the starting row, if necessary, to account for the row end pieces.

Starting in the corner, lay the first panel with the grooved side against the ¼" spacers. Slide the next panel into place and press-fit the groove of the second panel onto the tongue of the first. Check the edges against the wall.

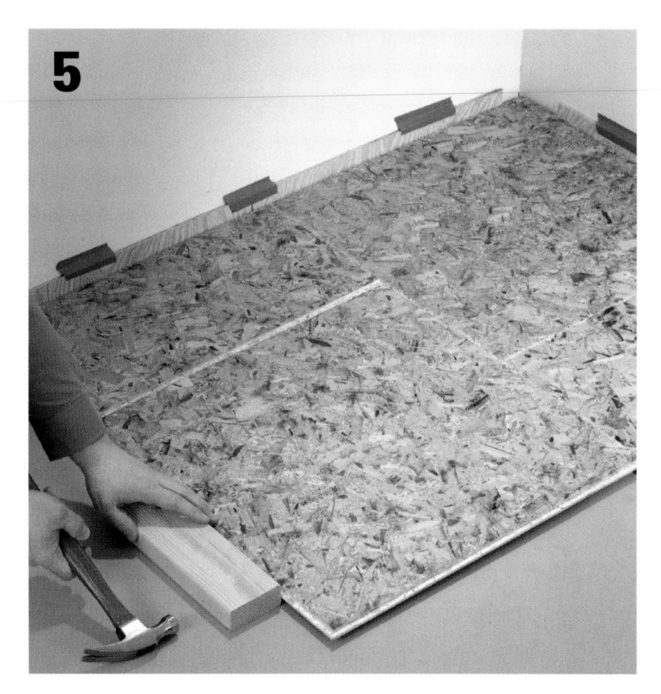

Repeat these steps to complete the first row. If necessary, tap the panels into place with a scrap piece of lumber and a rubber mallet or hammer—just be careful not to damage the tongue or groove edges. Starting with the second row, stagger the seams so that the panels interlock.

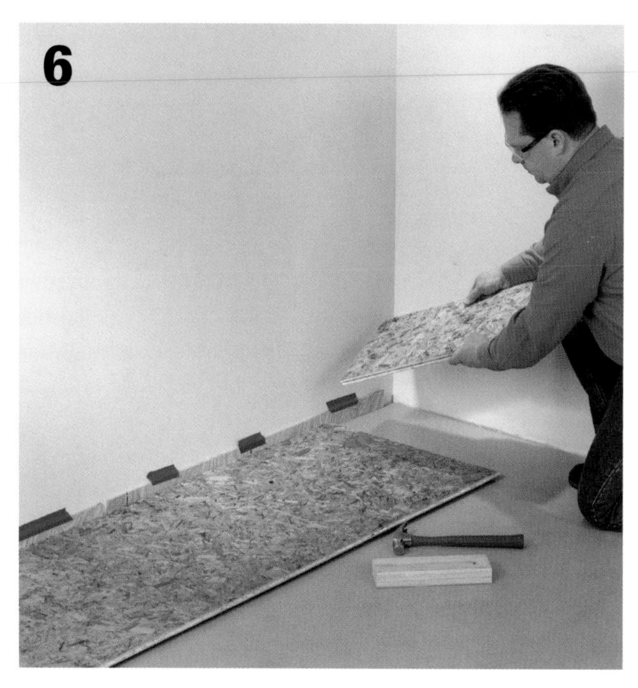

Cut the last panel to fit snugly between the next-to-last panel and the ¼" spacer on the far wall. Install the last panel at an angle and tap it down. Continue working from the starting point, checking after each row to be sure the panels are square and level.

When you reach the last row and last panel to complete your installation, you may have to cut the panel to fit. Measure for fit, allowing for the ¼" expansion gap from the wall. Cut the panel and fit it into place.

When all the panels are in place and the finished floor is installed, remove the spacers from around the perimeter of the room.

Preparing Basement Floors

How you prepare a concrete basement floor depends on the condition of the floor, the floor covering you plan to use, and how you want the floor to feel underfoot.

To lay flooring directly over concrete, prepare the floor to make it smooth and flat. Fill cracks, holes, and expansion joints with a vinyl or cement-based floor-patching compound. If the concrete is especially rough or uneven, apply a self-leveling, cement-based liquid floor leveler that fills low spots and dries to form a hard, smooth surface.

For a more resilient and comfortable basement floor, build a wood subfloor. It serves as a nailing surface for certain types of flooring. A subfloor does take up valuable headroom, so you may want to save space by using 1 × 4 sleepers instead of 2 × 4s. Consider how the added floor height will affect room transitions and the bottom step of the basement stairs.

Choose a floor covering before preparing a concrete floor, solve any moisture problems before installing a new surface, and be sure to follow manufacturer recommendations for installing on concrete. Both the results and the warranty may depend on it.

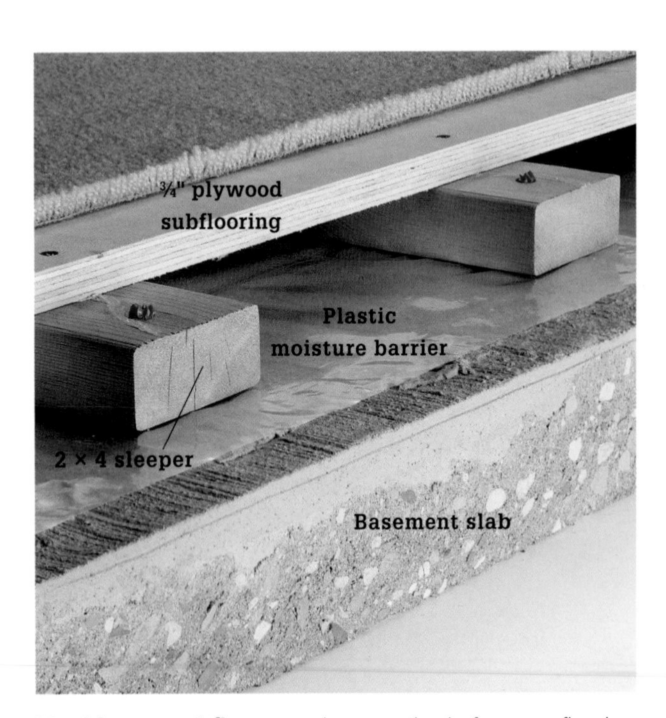

Most basement floors need preparation before new flooring can be laid. Patching compound and floor leveler can smooth rough concrete, while a wood subfloor creates a new surface that feels like a framed wood floor.

Tools & Materials ▸

- Vacuum
- Masonry chisel
- Hammer, trowel
- Floor scraper
- Long-nap paint roller
- Wheelbarrow
- Utility knife
- Gage rake
- 4-ft. level
- Circular saw
- Caulk gun
- Powder-actuated nailer
- Chalk line
- Drill
- Sledgehammer
- Vinyl floor-patching compound
- Concrete primer
- Floor leveler
- Pressure-treated 2 × 4s
- 6-mil polyethylene sheeting
- Packing tape
- Cedar shims
- Construction adhesive
- Concrete fasteners
- ¾" tongue-and-groove plywood
- 2" wallboard screws

Moisture Tip ▸

To test your floor for moisture, duct tape a 2 × 2 ft. piece of clear plastic to the concrete. Remove the plastic after 24 hours. If there's moisture on the plastic, you have a moisture problem. Do not install flooring until the problem has been fixed.

How to Patch Concrete Floors

Remove any loose or flaking concrete and vacuum the floor with a masonry chisel and hammer. Mix a batch of vinyl floor-patching compound, following the manufacturer's directions. Apply the compound using a smooth trowel, slightly overfilling the cavity. Smooth the patch flush with the surrounding surface.

After the compound has cured, use a floor scraper to scrape the patched areas smooth.

How to Apply Floor Leveler

Remove any loose material and clean the concrete thoroughly so it's free of dust, dirt, oils, and paint. Apply an even layer of concrete primer to the entire surface using a long-nap paint roller. Let the primer dry completely before continuing.

Following the manufacturer's instructions, mix the floor leveler with water. The batch should be large enough to cover the entire floor area to the desired thickness (up to 1"). Pour the leveler over the floor.

Distribute the leveler evenly, using a gage rake or spreader. Work quickly since the leveler begins to harden in 15 minutes. Use a trowel to feather the edges and create smooth transitions with uncovered areas. Let the leveler dry for 24 hours.

How to Build a Basement Subfloor

Chip away loose or protruding concrete with a masonry chisel and hammer, then vacuum the floor. Roll out strips of 6-mil polyethylene sheeting, extending them 31" up each wall, Overlap strips by 6", then seal the seams with packing tape. Temporarily tape the edges along the walls. Be careful not to damage the sheeting.

Lay out pressure-treated 2 x 4s along the perimeter of the room. Position the boards ½" in from all walls (inset). *NOTE: Before laying out the sleepers, determine where the partition walls will go. If a wall will fall between parallel sleepers, add an extra sleeper to support the planned wall.*

Using a circular saw, cut the sleepers to fit between the perimeter boards, leaving a ¼" gap at each end. Position the first sleeper so its center is 16" from the outside edge of the perimeter board. Lay out the remaining sleepers, using 16"-on-center spacing.

Where necessary, use tapered cedar shims to compensate for dips and variations in the floor. Place a 4-ft. level across neighboring sleepers. Apply construction adhesive to two wood shims. Slide the shims under the board from opposite sides until the board is level with adjacent sleepers.

Fasten the perimeter boards and sleepers to the floor using a powder-actuated nailer or masonry screws. Drive a fastener through the center of each board at 16" intervals. Fastener heads should not protrude above the board's surface. Place a fastener at each shim location, making sure the fastener penetrates both shims.

Establish a control line for the first row of plywood sheets by measuring 49" from the wall and marking the outside sleeper at each end of the room. Snap a chalk line across the sleepers at the marks. Run a ¼"-wide bead of adhesive along the first six sleepers, stopping just short of the control line.

Position the first sheet of plywood so the end is ½" away from the wall and the grooved edge is flush with the control line. Fasten the sheet to the sleepers using 2" wallboard screws. Drive a screw every 6" along the edges and every 8" in the field. Don't drive screws along the grooved edge until the next row of sheeting is in place.

Install the remaining sheets in the first row, maintaining an ⅛" gap between ends. Begin the second row with a half-sheet (4 ft. long) so the end joints between rows are staggered. Fit the tongue of the half sheet into the groove of the adjoining sheet. If necessary, use a sledgehammer and wood block to help close the joint. After completing the second row, begin the third row with a full sheet. Alternate this pattern until the subfloor is complete.

Building Attic Floors

Existing floors in most unfinished attics are merely ceiling joists and are too small to support living spaces. However, if your floor already has floor trusses, joists 2 × 8 or larger, or the same framing as the floor on your main level, it probably doesn't need additional framing.

Before you build, consult an architect, engineer, or general contractor as well as a local building inspector. Ask what size joists you'll need and which options your local building codes allow. Joist sizing is based on joist span and spacing; an attic floor must be able to support 40 pounds per square feet (psf) of live load, such as occupants and furniture, and 10 psf of dead load, including wallboard and floor covering.

The simplest method for strengthening your attic floor is to install an additional joist next to each existing joist and nail the two together. This process, known as sistering, is done when the current joists are damaged or loose, squeak, or can't support additional weight. This method only works for joists that are 2 × 6 or larger, closely spaced, and without obstructions.

An alternative is to build a new floor by placing larger joists between the existing ones. By resting the joists on 2 × 4 spacers, you avoid obstructions and minimize damage to the ceiling surfaces below. Be aware that the spacers will reduce your headroom by 1½" in addition to the depth of the joists.

Floor joist cavities offer space for concealing the plumbing, wiring, and ductwork servicing your attic, so consider these systems as you plan. Plan the locations of partition walls to determine if additional blocking between joists is necessary.

When the framing is done, the mechanical elements and insulation are in place, and everything has been inspected and approved, complete the floor by installing ¾" tongue-and-groove plywood. If your remodel will include kneewalls, you can omit the subflooring behind the kneewalls, but a complete subfloor adds strength and provides a sturdy surface for storage.

Tools & Materials ▸

- Circular saw
- Rafter square
- Drill
- Tape measure
- Caulk gun
- Joist lumber
- 16d common nails
- 10d common nails
- 8d common nails
- 2 × 4 lumber
- ¾" tongue-&-grove plywood
- Construction adhesive
- 2¼" wallboard screws

How to Add Sister Joists

Remove all insulation from the joist cavities and carefully remove any blocking or bridging between the joists. Determine the lengths for the sister joists by measuring the existing joists. Also measure the outside end of each joist to determine how much of the top corner needs to be cut away to fit the joist beneath the roof sheathing. *NOTE: Joists that rest on a bearing wall should overlap each other by at least 3".*

Before cutting the joists, sight down both narrow edges of each board to check for crowning, which is an upward arching along the length of the board. Draw an arrow pointing toward the arch. Joists must be installed crown side up. Cut the board to length, then clip the top outside corner to match the existing joist.

Set the sister joists in place, flush against the existing joists with their ends aligned. Toenail each sister joist to the top plates of both supporting walls using two 16d common nails.

Face-nail the joists together using 10d common nails. Drive three nails in a row, spacing the rows 12" to 16" apart. To minimize damage (such as cracking and nail popping) to the ceiling surface below, you can use a pneumatic nail gun or 3" lag screws instead of nails. Install new blocking between the sistered joists as required by the local building code.

How to Build an Attic Floor Using New Joists

Remove any blocking or bridging from between the existing joists, being careful not to disturb the ceiling below. Cut 2 × 4 spacers to fit snugly between each pair of joists. Lay the spacers flat on the top plate of each supporting wall. Nail the spacers to the top plates using 16d common nails.

Create a layout for the new joists by measuring across the tops of existing joists and using a rafter square to transfer the measurements down to the spacers. Using 16"-on-center spacing, mark the layout along one exterior wall, then mark an identical layout on the interior bearing wall. The layout on the opposing exterior wall will be offset 1½" to account for the joist overlap at the interior wall.

Measure from the outer edge of the exterior wall to the far edge of the interior bearing wall. The joists must overlap above the interior wall by at least 3". Measure the outside end of each joist to determine how much of the top corner needs to be cut away to fit under the roof sheathing. Cut the joists to length, then clip the top outside corners as necessary.

Set the joists in place on their layout marks. Toenail the outside end of each joist to the spacer on the exterior wall using three 8d common nails.

Nail the joists together where they overlap atop the interior bearing wall, using three 10d common nails for each connection. Toenail the joists to the spacers on the interior bearing wall, using 8d common nails.

Install blocking or bridging between the joists, as required by your local building code. As a suggested minimum, the new joists should be blocked as close as possible to the outside ends and at the points where they overlap at the interior wall.

How to Install Subflooring

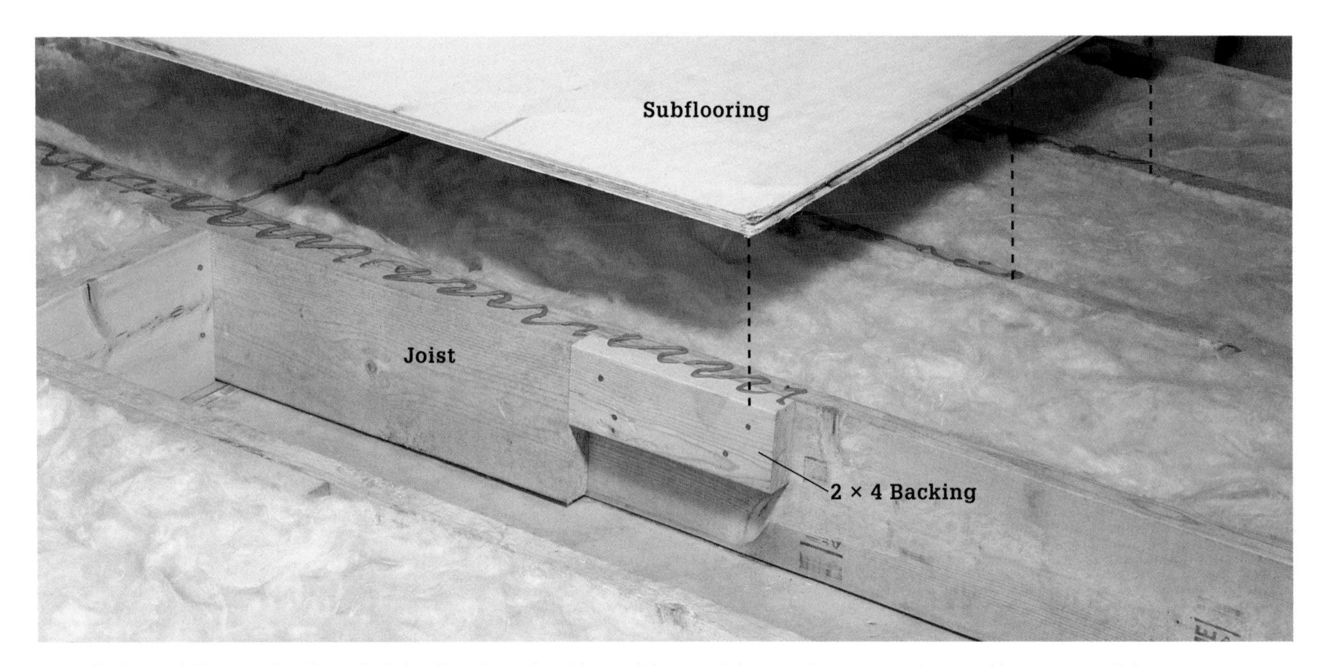

Install the subfloor only after all of the framing, plumbing, wiring, and ductwork are complete and have passed the necessary building inspections. Install insulation as needed and complete any caulking necessary for soundproofing. Fasten the subflooring sheets with construction adhesive and 2¼" wallboard or deck screws, making sure the sheets are perpendicular to the joists and the end joints are staggered between rows. Where joists overlap at an interior bearing wall, add backing as needed to compensate for the offset in the layout. Nail a 2 × 4 (or wider) board to the face of each joist to support the edges of the sheets.

Installing a Floor-warming System

To enjoy the look of ceramic tile without the discomfort of cold feet, install a floor-warming system. A typical system uses an electric wire mesh that heats up when energized, like an electric blanket. These systems are designed to heat only the floor, so they require little energy. The mesh is installed under the floor covering and hard-wired to a 120-volt GFCI circuit. A thermostat and timer control the system automatically.

For a successful project, check the resistance often as you install the wire heating mesh, and make sure you have adequate power for the circuit that controls the in-floor heat. If you're installing a new circuit, consider hiring an electrician to make the connection at the service panel. To order a floor-warming system, contact a manufacturer or dealer.

Floor-warming systems can also be used under laminate, vinyl, and floating floors. They are not suited for use under a wood covering that requires nailing, because nails can damage the wires. Also, use rosin paper instead of asphalt felt paper as an underlayment. When asphalt paper warms up, it can smell very unpleasant.

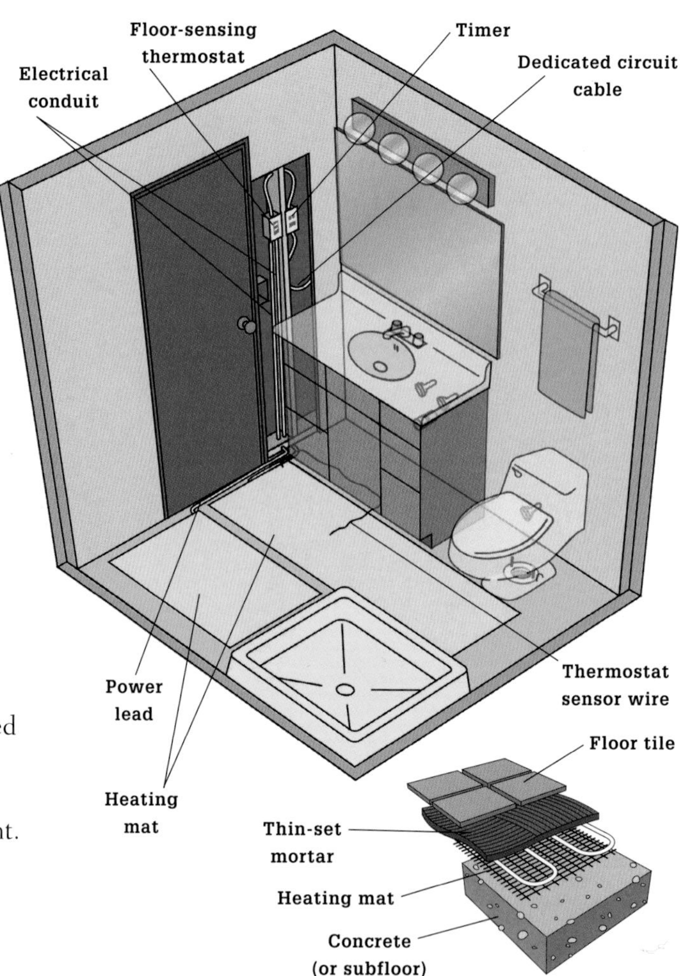

Tools & Materials ▸

- Multi-tester
- Drill
- Plumb bob
- Chisel
- Tubing cutter
- Combination tool
- Vacuum
- Chalk line
- Grinder
- Glue gun
- Fish tape
- Aviation snips
- ⅜ × ¼" square-notched trowel
- Tile tools and materials
- Floor-warming system
- 2½ × 4" double-gang electrical box with 4" adapter cover
- 2½"-deep single-gang electrical box
- ½"-dia. thin-wall conduit
- Setscrew fittings
- 12-gauge NM cable
- Cable clamps
- Double-sided tape
- Electrical tape
- Insulated cable clamps
- Wire connectors

Tip ▸

Floor-warming systems must be installed on a circuit with adequate amperage and a GFCI breaker (some systems have built-in GFCIs). Smaller systems may tie into an existing circuit, but larger ones often need a dedicated circuit. Follow all local building and electrical codes that apply to your project.

How to Install a Floor-warming System

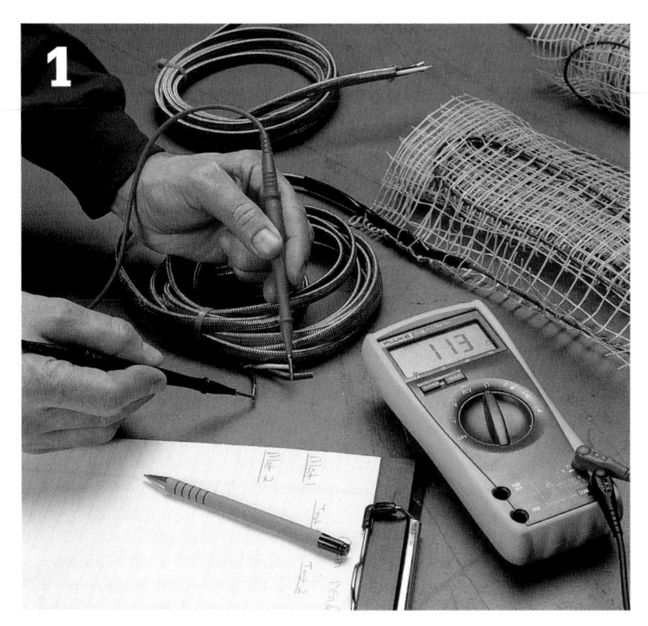

Check the resistance value (ohm) of each heating mat using a digital multi-tester. Record the reading. Compare your reading to the factory-tested reading noted by the manufacturer. Your reading must fall within the acceptable range determined by the manufacturer. If it doesn't, the mat has been damaged and should not be installed. Contact the manufacturer for assistance.

Remove the wall surface to expose the framing. Locate the electrical boxes approximately 60" from the floor, making sure the power leads on the heating mats will reach the electrical box. Mount a 2 × 2"-deep × 4"-wide double-gang electrical box for the thermostat to the wall stud. Mount a single-gang electrical box for the timer on the other side of the stud.

Use a plumb bob or level to mark points on the bottom wall plate directly below the two knockouts on the thermostat box. At each mark, drill a ½" hole through the top of the plate. Drill two more holes as close as possible to the floor through the side of the plate, intersecting the top holes. Clean up the holes with a chisel to ensure smooth routing.

Cut two lengths of thin-wall electrical conduit with a tubing cutter to fit between the thermostat box and the bottom plate. Place the bottom end of each conduit about ¼" into the respective holes in the bottom plate and fasten the top ends to the thermostat box using setscrew fittings. If you're installing three or more mats, use ¾" conduit instead of ½".

(continued)

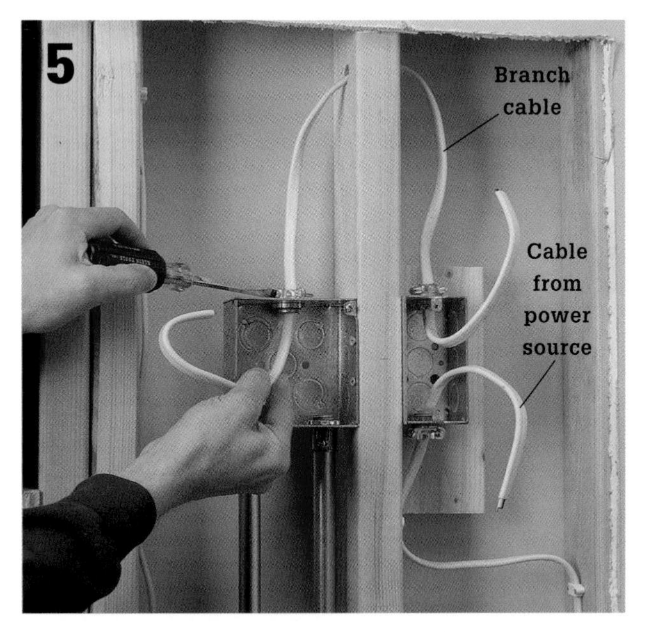

Run 12-gauge NM electrical cable from the service panel (power source) to the timer box. Attach the cable to the box with a cable clamp, leaving 8" of extra cable extending into the box. Drill a ⅝" hole through the center of the stud about 12" above the boxes. Run a short branch cable from the timer box to the thermostat box, securing both ends with clamps. The branch cable should make a smooth curve where it passes through the stud.

Vacuum the floor thoroughly. Plan the ceramic tile layout and snap reference lines for the tile installation (see page 80). Spread the heating mats over the floor so the power leads are close to the electrical boxes. Position the mats 3" to 6" away from walls, showers, bathtubs, and toilet flanges. Place the mats in the kick space of a vanity, but not under the vanity cabinet or over expansion joints in the concrete slab. Set the edges of the mats close together, but don't overlap them. The heating wires in one mat must be at least 2" away from the wires in the neighboring mat.

Confirm that the power leads still reach the thermostat box. Secure the mats to the floor using strips of double-sided tape spaced every 2 ft. Make sure the mats are lying flat with no wrinkles or ripples. Press firmly to secure the mats to the tape.

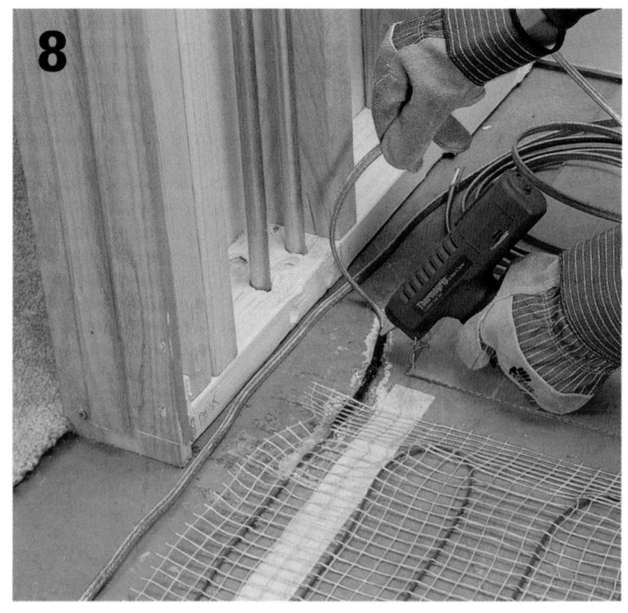

Create recesses in the floor for the connections between power leads and heating-mat wires, using a grinder or a cold chisel and hammer. These insulated connections are too thick to lay under the tile and must be recessed to within ⅛" of the floor. Clean away any debris and secure the connections in the recesses with a bead of hot glue.

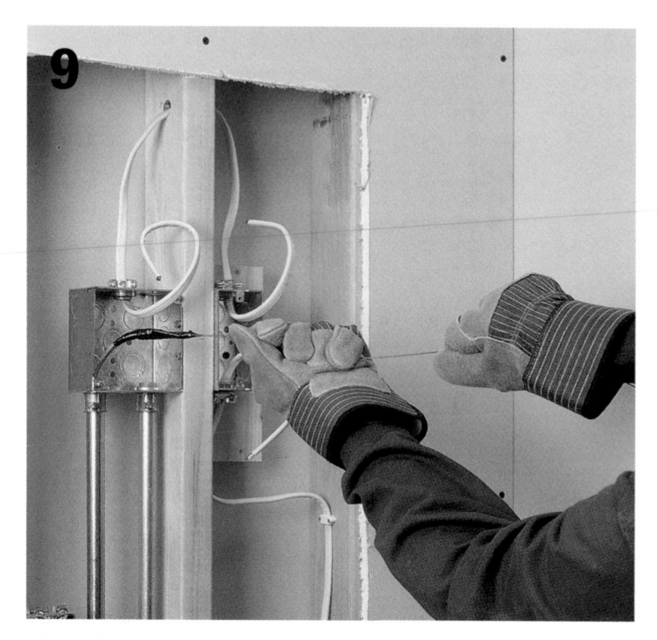

Thread a steel fish tape down one of the conduits and attach the ends of the power leads to the fish tape using electrical tape. Pull the fish tape and leads up through the conduit. Disconnect the fish tape, then secure the leads to the box with insulated cable clamps. Use aviation snips or linesman's pliers to cut off excess from the leads, leaving 8" extending past the clamps.

Feed the heat sensor wire through the remaining conduit and weave it into the mesh of the nearest mat. Use dabs of hot glue to secure the sensor wire directly between two blue resistance wires, extending it 6" to 12" into the mat. Test the resistance of the heating mats with a multi-tester as you did in step 1 to make sure the resistance wires have not been damaged. Record the reading.

Install the floor tile as shown on pages 150 to 155. Using thin-set mortar as an adhesive, spread it carefully over the floor and mats with a ⅜" × ¼" square-notched trowel. Check the resistance of the mats periodically as you install the tile. If a mat becomes damaged, clean up any exposed mortar and contact the manufacturer. When the installation is complete, check the resistance of the mats once again and record the reading.

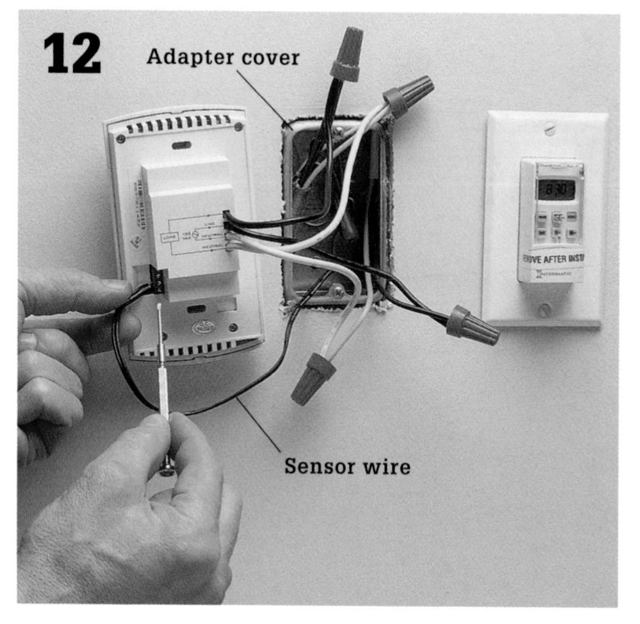

Install an adapter cover (mud ring) to the thermostat box, then patch the wall opening with wallboard. Complete the wiring connections for the thermostat and timer following the manufacturer's instructions. Attach the sensor wire to the thermostat setscrew connection. Apply the manufacturer's wiring labels to the thermostat box and service panel. Mount the thermostat and timer. Complete the circuit connection at the service panel or branch connection. After the flooring materials have cured, test the system.

Installing an In-floor Electrical Outlet

Installing an electrical outlet in the floor is surprisingly simple. Floor box kits have everything you need to add an outlet in any room. The outlet is covered when not in use to protect against debris falling into the receptacle.

The outlet is placed in the subfloor before the floor covering is installed. If the floor is already finished, remove an area of flooring.

Before installing an outlet, check with your local building department for any restrictions.

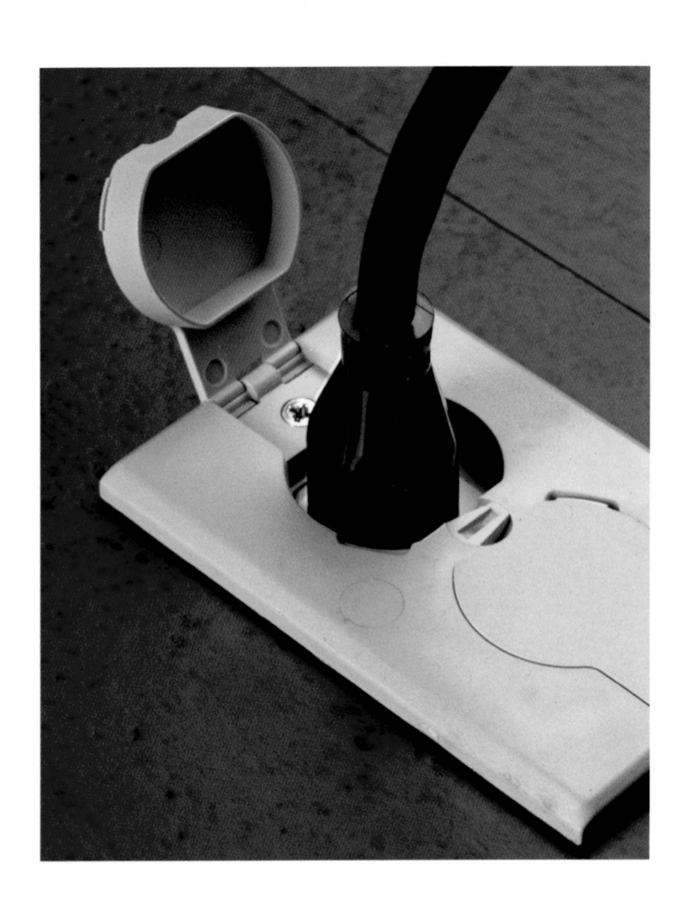

Tools & Materials ▸

Floor box assembly kit
Screwdriver
Cordless drill
Jig saw

How to Install an In-floor Electrical Outlet

Place the box on the subfloor where the outlet will be installed. Make sure it sits next to a floor joist. Trace around the box. Remove the box, then drill holes at the corners on the floor. Cut out the opening, using a jig saw.

Place the clip for the floor box in the opening so the lip sits on top of the subfloor. Attach the clip to the subfloor and joist using the four 1¼" self-tapping screws that came with the kit.

Slide the box onto the clip so the adjusting screw is aligned with the threads on the clip. Screw the box into the clip, lowering the box into place. Do not set the box all the way to the subfloor.

Turn off the power at the main power source. Insert electrical wires into the box and wire the receptacle in accordance with your local building codes. Push the wires and receptacle inside the box.

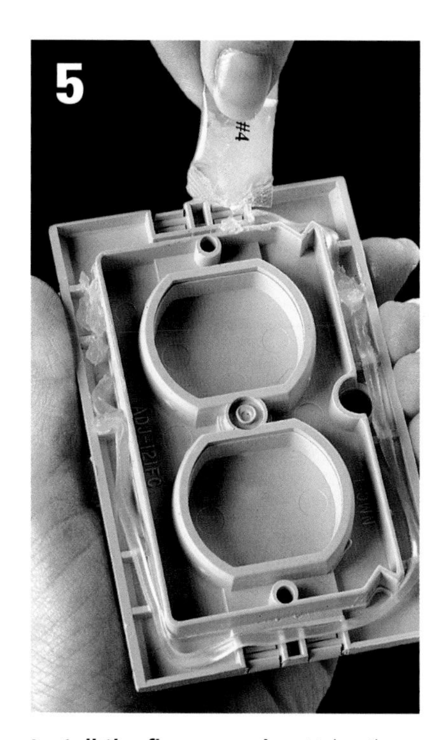

Install the floor covering. Using the sealant that came with the kit, apply a ⅛"-thick bead of sealant around the outside rib of the cover where it fits over the box.

With the outlet caps open, set the cover on the box. Align the holes with the recessed bosses and the button hole with the adjustable screw. Insert two machine screws and tighten, but don't overtighten.

Turn the adjustable screw to set the face of the outlet at the desired height. Place the button that came with the kit in the button hole.

Chinese Patent No. ZL991178092
Mexican Patent No. 204281
Teragren commercial
SYNERGY
Teragren commercial
SYNERGY
STRAND BAMBOO FLOORING
Strand Bamboo Flooring
STRAND BAMBOO FLOORING
STRAND BAMBOO FLOORING

Installations

Of all the home improvements you can choose to handle yourself, flooring may be the most rewarding. You probably own and know how to use most of the tools, and the others are easy to get from rental centers. If you're concerned about your ability, take courage. Installing a floor does not require exceptional strength or skill—just planning and care.

In this Chapter:

Getting Started

Your first row of flooring, your first few tiles, or your first piece of sheeting sets the direction for the rest of your floor. That means you need to create a starting point carefully, or the whole project may run off course. You can do this by carefully planning your layout and establishing accurate reference lines.

In general, tile flooring begins at the center of the room and is installed in quadrants along layout lines, also called working lines. After establishing reference lines that mark the center of the room, lay the tile in a dry run along those lines to ensure you won't have to cut off more than half of a tile in the last row. If necessary, adjust your reference lines by half the width of the tile to form your layout lines.

For most floating floors and tongue-and-groove floors, you only need a single reference line along the starting wall. If your wall is straight, you don't even need a working line. You can place spacers along the wall and butt the first row of flooring against the spacers. However, this only works if your wall is straight. If it is bowed or out of square, your layout will be affected.

To check your reference lines for squareness, use the 3-4-5 triangle method. Measuring from your centerpoint, make a mark along a reference line at 3 ft. and along a perpendicular reference line at 4 ft. The distance between the two points should be exactly 5 ft. If it's not, adjust your lines accordingly.

Tools & Materials ▸

- Tape measure
- Chalk line
- Framing square
- Hammer
- 8d finish nails
- Spacers

How to Establish Reference Lines for Tile

Mark the centerpoint of opposite walls, then snap a chalk line between the marks. Mark the centerpoint of the chalk line. Place a framing square at the centerpoint so one side is flush with the chalk line. Snap a perpendicular reference along the adjacent side of the framing square.

Snap chalk lines between the centerpoints of opposite walls to establish perpendicular reference lines. Check the lines for squareness using the 3-4-5 triangle method.

How to Establish Reference Lines for Wood and Floating Floors

If your wall is out of square or bowed, make a mark on the floor ½" from the wall at both ends and snap a chalk line. Drive 8d finish nails every 2" to 3" along the line. Use this as your reference line and butt the first row of flooring against the nails (see page 94).

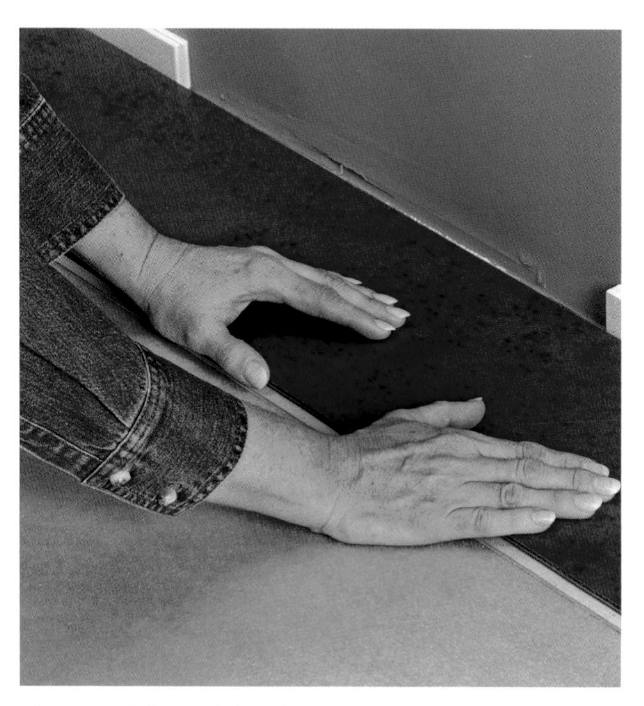

If your wall is straight, place 12" spacers along the wall, then butt your flooring up against the spacers.

How to Check the Walls for Square

If the flooring you have chosen depends on straight edges, whether because of the material or its pattern, you will be better prepared for variations if you know about them before you begin. Use these four simple techniques to find out if the corners in your room are square, and whether the walls are straight.

To check any corner for square over a longer distance than the carpenter's square can reach, use the 3-4-5- method. Measure and mark a point 3 feet out from the corner in one direction. Do the same on the adjacent wall at a point 4 feet from the corner; make sure the two points are the same height from the floor. Then measure between the two points. If this third measurement is exactly 5 feet, the corner is square.

To find whether a wall bulges in to the room or bows out, you can use a mason's line and a measuring tape. Measure out an equal distance from the corners at two ends of a wall, and fasten a mason's line between the two points. Then measure the distance between the mason's line and the wall at several points. This will reveal any bulges or bows in the wall.

Tip ▸

Hold a carpenter's square perfectly level, as it would be on the floor, and fit it into the corner. If the point of the square does not touch exactly where the walls meet, the corner has an acute angle. If the corner of the square touches but the two ends do not touch both walls at the same time, the corner has an obtuse angle.

Tip ▸

With a helper, stretch a measuring tape across the room diagonally, from one corner to the opposite. Record the measurement, then repeat the measurement between the other two corners. If the measurements are the same, the corners are more or less square. *Note: If the diagonal measurements do not match, one or more of the corners is probably not square. Before you can compensate for the difference, you will need to know whether the individual corners are acute (less than 90°) or obtuse (more than 90°). See Tip to left.*

Cutting Door Casing

Unless you plan to install carpet, you will want your floor covering to fit under your door casing. This allows the casing to cover the gap between the flooring and the wall, and it allows wood floors to expand and contract without dislodging the casing. If you try to butt the flooring against the casing, chances are you'll end up with an unsightly gap.

It only takes a few minutes to cut the casing. If you're installing ceramic tile or parquet, keep in mind you'll be placing the flooring over adhesive, so cut the casing about an ⅛" above the top of tile to allow for the height of the adhesive.

These directions show the casing being cut to accommodate ceramic tile. Because the tile will be placed on top of cementboard, a piece of cementboard is placed under the tile when the casing is marked.

Tools & Materials ▸

Jamb saw

Floor covering

How to Cut Door Casing

Place a piece of flooring and underlayment against the door casing. Mark the casing about an ⅛" above the top of the flooring.

Cut the casing at the mark using a jamb saw.

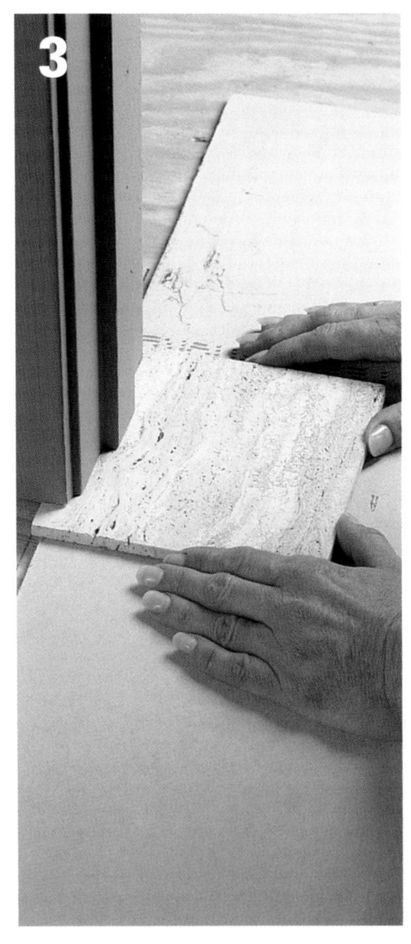

Slide a piece of flooring under the door jamb to make sure it fits easily.

Thresholds & Moldings

When you install wood or laminate floors, leave a ½" gap between the perimeter of the floor and the walls to allow the wood to expand and contract with changes in temperature and humidity.

You will also have gaps that need to be covered at thresholds, between rooms, and around small obstacles, such as pipes. For every situation, there is a molding to fit your needs.

A floor isn't truly finished until all of the pieces are in place. These moldings help give your floors a professional look. The names for moldings may differ slightly between manufacturers.

Wood molding is used for a smooth transition between the hardwood in the dining area and the tile in the adjoining room.

A. Carpet reducers are used to finish off and create a smooth transition between flooring and carpeting.
B. Stair nosing is used to cover the exposed edges of stairs where the risers meet the steps. It is also used between step-downs and landings.
C. Baby threshold is used in place of baseboards and quarter round in front of sliding glass doors or door thresholds, to fill the gap between the floor and door.
D. Reducer strips, also called transition strips, are used between rooms when the floors are at different heights and composed of different materials.
E. Overlap reducers are also used between rooms when one floor is at a different height than an adjoining room.
F. T-moldings are used to connect two floors of equal height. They are also used in doorways and thresholds to provide a smooth transition. T-moldings do not butt up against the flooring, allowing the wood to expand and contract under it.
G. Baseboards are used for almost all types of floors and are available in a wide variety of designs and thicknesses. They are applied at the bottom of walls to cover the gap between the floor and walls.
H. Quarter round, similar to shoemolding, is installed along the bottom edge of base board and sits on top of the floor. It covers any remaining gaps between the floor and walls.

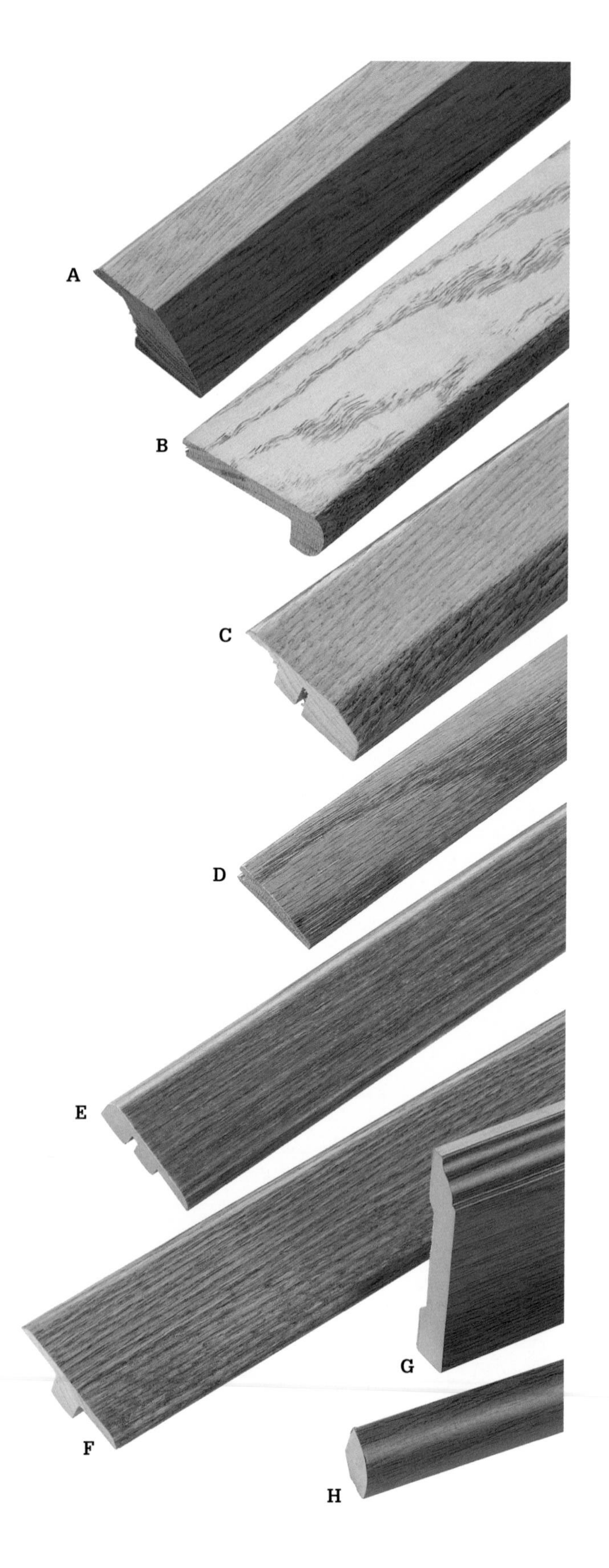

Hardwood

Hardwood is a traditional favorite for floors. It adds character and sets the mood for a room. The classic beauty of hardwood lends itself to any decorating style and trend while providing a consistent element for tying rooms together.

Although hardwood was once reserved for formal rooms, it's now used in virtually every room in the home, including kitchens. Unlike other floor coverings, hardwood will last a lifetime. If properly maintained, it can actually look more charming as it ages.

As you'll see in this section, there are a variety of wood floor coverings that require different installation techniques. Tongue-and-groove strip flooring is installed using a power nailer, parquet and end grain floors are set in adhesive, while floating floors fasten together at the tongue-and-groove connections and are not connected to the floor at all.

Wood floors absorb moisture from the humidity in the air, causing the wood to expand. When the air is dry, wood contracts. The flooring is kept ½" from the walls to allow for this expansion and contraction. The gap is covered by the baseboard and base shoe.

Wood must be "acclimated" to the room in which it will be installed. Place the flooring in the room under normal temperatures and humidity conditions. The length of this acclimation period varies, sometimes taking up to a full week, so check manufacturer's recommendations before installing.

Tools for Hardwood Floors

Tools and Materials for hardwood include: Power tools for hardwood flooring installation include: miter saw (A), circular saw (B), jig saw (C), rubber mallet (D), power nailer (E), cordless drill (F).

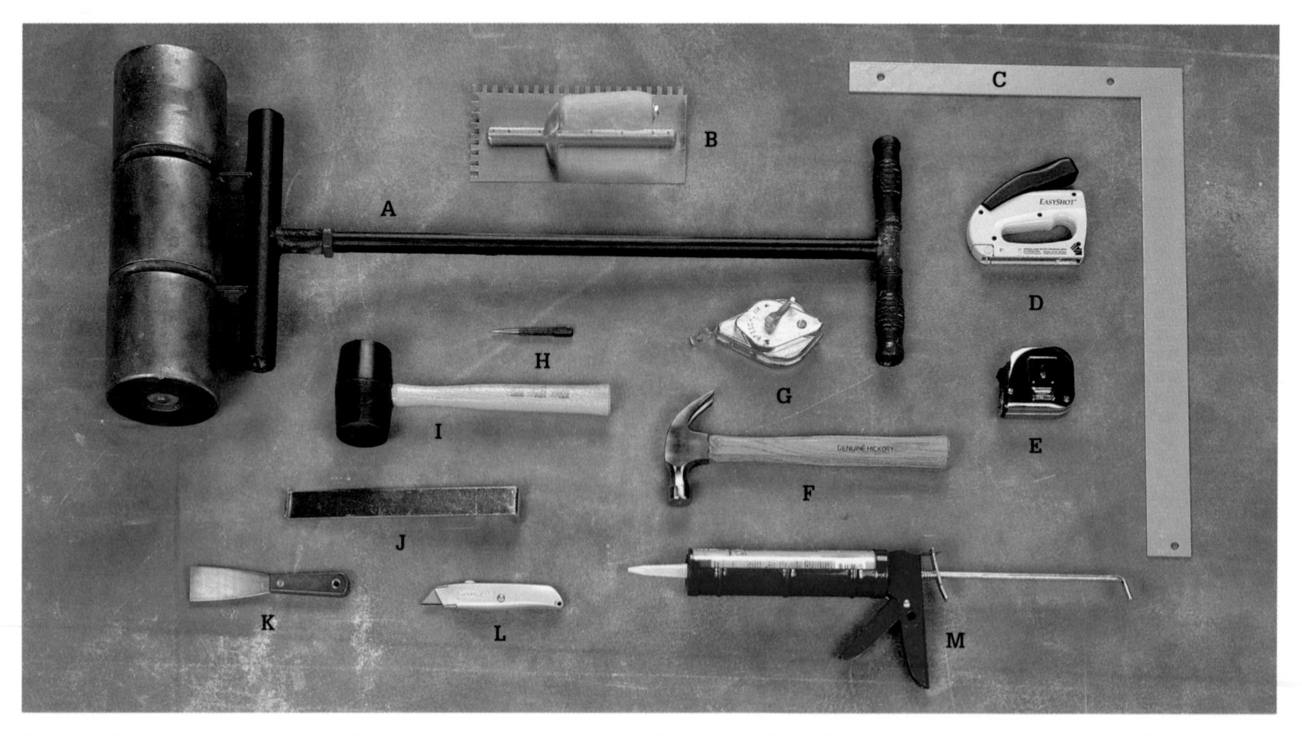

Hand tools for hardwood flooring installation include: floor roller (A), notched trowel (B), framing square (C), stapler (D), tape measure (E), hammer (F), chalk line (G), nail set (H), rubber mallet (I), floor pull bar (J), putty knife (K), utility knife (L), caulk gun (M).

How to Cut Hardwood Flooring

Ripcut hardwood planks from the back side to avoid splintering the top surface. Measure the distance from the wall to the edge of the last board installed, subtracting ½" to allow for an expansion gap. Transfer the measurement to the back of the flooring, and mark the cut with a chalk line.

When ripcutting hardwood flooring with a circular saw, place another piece of flooring next to the one marked for cutting to provide a stable surface for the foot of the saw. Clamp a cutting guide to the planks to ensure a straight cut.

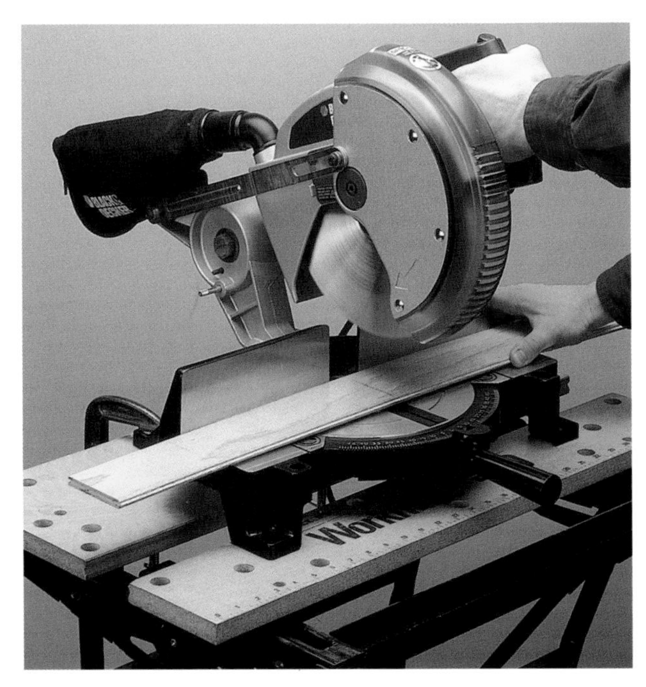

Crosscut hardwood flooring using a power miter box. Place the top surface face up to prevent splintering.

Make notched or curved cuts in hardwood flooring with a coping saw or jig saw. If using a jig saw, the finished surface should face down. Clamp the flooring to your work surface when cutting.

Installing Tongue-and-Groove Hardwood Flooring

Tongue-and-groove hardwood flooring has always been popular with homeowners. It offers an attractive look, is one of the longest lasting floor coverings, and can be stripped and refinished to look like new.

Oak has been the most common type of strip flooring because of its durability and wood graining, and it's the species most people think of when hardwood is mentioned. Other woods, such as maple, cherry, and birch, are also becoming popular.

Exotic species of wood from around the world are now finding their way into American homes as people want a premium strip or plank floor that is unique and stylish, and expresses their personalities. The more than 60 exotic hardwoods include Brazilian cherry, Australian cypress, Honduran mahogany, tobaccowood, teak, zebrawood, and bamboo—which is not really wood but a type of grass.

This section describes how to install nailed-down tongue-and-groove flooring, how to install a decorative medallion, and how to install tongue-and-groove strip flooring over troweled-on adhesive. Customizing your floor with borders, accents, and medallions is easier than you think. A number of manufacturers produce a variety of decorative options made to match the thickness of your floor.

How to Install Tongue-and-Groove Hardwood Flooring

Cover the entire subfloor with rosin paper. Staple the paper to the subfloor, overlapping edges by 4". Cut the paper with a utility knife to butt against the walls.

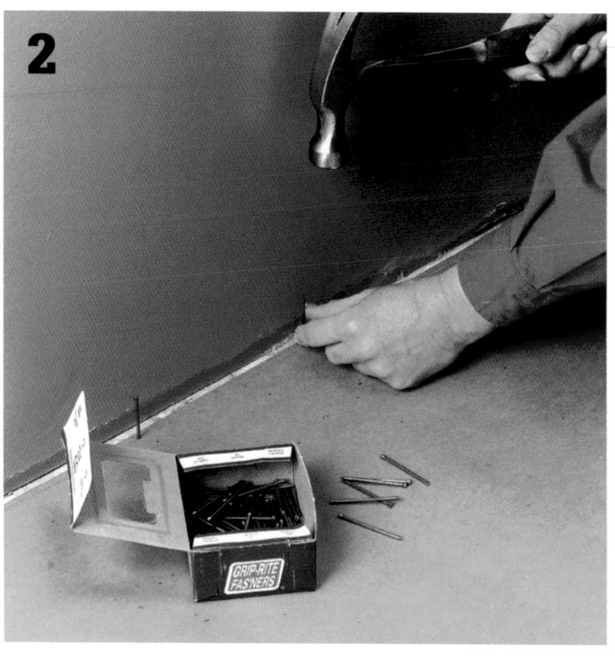

Make a mark on the floor ½" from the starter wall at both ends of the wall. Snap a chalk line between the marks. Nail 8d finish nails every 2" to 3" along the chalk line to mark the location for your first row.

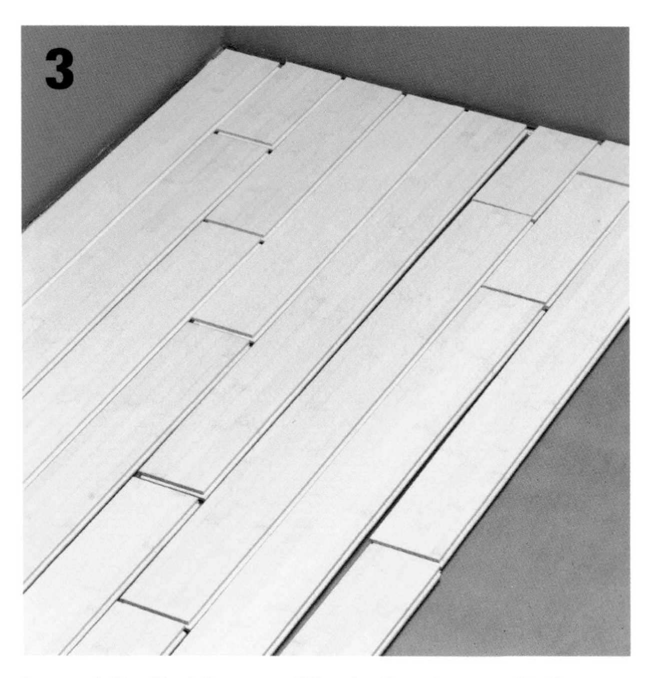

Lay out the first 8 rows of flooring in a dry run with the groove side facing the wall. Make sure the first row of boards is straight. Arrange the boards to get a good color and grain mix. Offset the ends by at least 6".

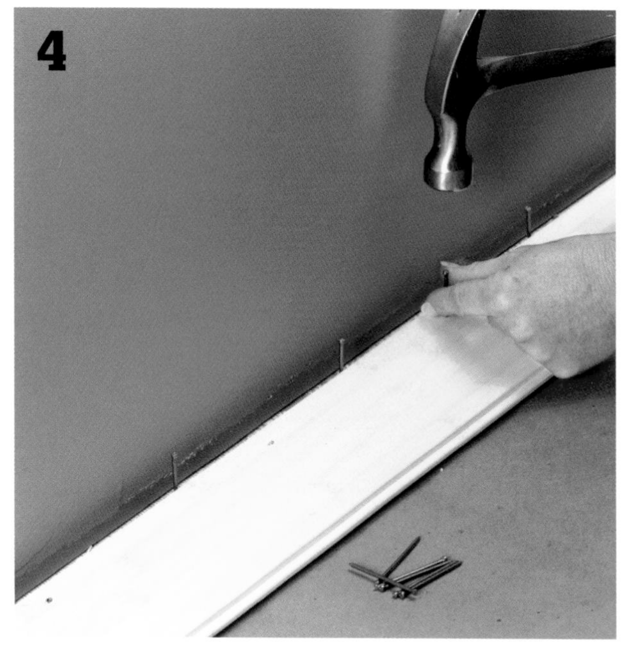

Place the starter row against the nails on the chalk line. Drill pilot holes in the flooring every 6" to 8", about ½" from the groove edge. Face nail the first row until the nail heads are just above the boards, then sink them using a nail set. (Be careful not to hit the boards with your hammer or you'll mar the surface.)

(continued)

Drill pilot holes every 6" to 8" directly above the tongue, keeping the drill at a 45° angle.

Tip ▸

To install crooked boards, drill pilot holes above the tongue and insert nails. Fasten a scrap board to the subfloor using screws. Force the floor board straight using a pry bar and a scrap board placed in front of the flooring. With pressure on the floor board, blind nail it into place.

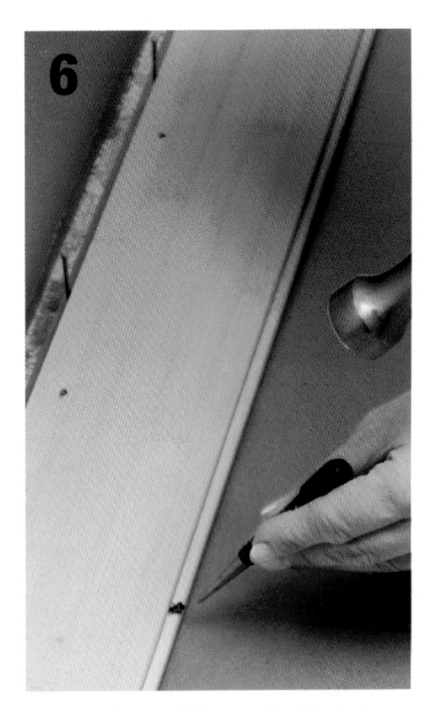

Blind nail a nail into each pilot hole. Keep the nail heads ½" out, then set them just below the surface, using a nail set.

Set the second row of boards in place against the starter row, fitting together the tongue and groove connections. Use a scrap board and rubber mallet to tap the floor boards together. Drill pilot holes and blind nail the boards. Do this for the next few rows.

To install the last board in a row, place the tongue and groove joints together, then place a flooring pull bar over the end of the board. Hit the end of the pull bar with a hammer until the board slides into place. Stay ½" away from the walls.

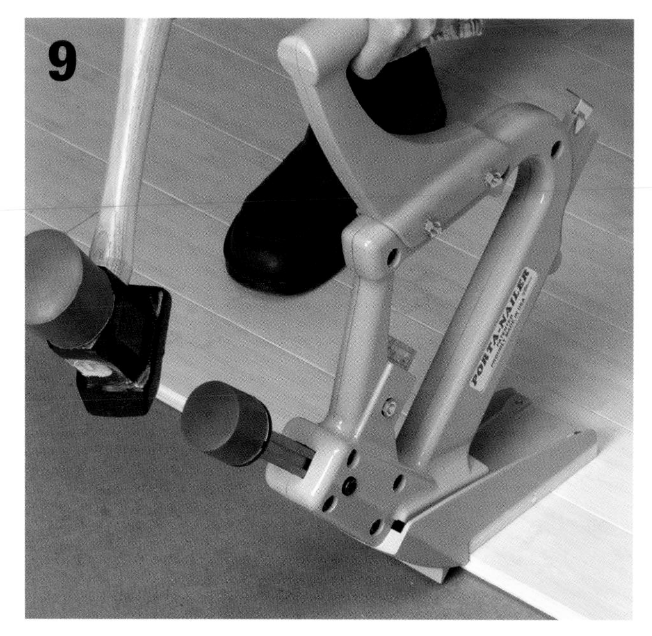

Once you have several rows installed and have enough room, use a power nailer. Place the nailer lip over the edge of the board and strike it with a rubber mallet. Drive a nail 2" from the end of each board and about every 8" in the field. Keep a few rows of flooring laid out ahead of you as you work, and keep the joints staggered.

When you're out of room for the power nailer, drill pilot holes and blind nail the boards. For the last rows, drill pilot holes in the top of the boards, ½" from the tongue, and face nail them. The last row may need to be ripped to size. Pull the last row into place using the flooring pull bar, leaving a ½" gap along the wall. Drill pilot holes and face nail.

Install a reducer strip or transition strip between the wood floor and an adjoining room. Cut the strip to size. Fit the strip's groove over the floor board's tongue, then drill pilot holes and face nail. Set the nails with a nail set. Fill all visible nail holes with wood putty.

Tip ▸

To install around an object, cut a notch in the board. For larger obstacles, cut 45° miters in boards so the grooves face away from the object. Rip the tongues off the boards. Set the boards against the object and the flooring, fitting the mitered ends together. Drill pilot holes and face nail in place. Apply silicone caulk between the floor.

To reverse directions of the tongue and groove for doorways, glue a spline into the groove of the board. Fit the groove of the next board onto the spline, then nail the board in place.

How to Install Tongue-and-Groove Flooring Using Adhesive

To establish a straight layout line, snap a chalk line parallel to the longest wall, about 30" from the wall. Kneel in this space to begin flooring installation.

Apply flooring adhesive to the subfloor on the other side of the layout line with a notched trowel, according to the manufacturer's directions. Take care not to obscure the layout line with adhesive.

Apply wood glue to the grooved end of each piece as you install it to help joints stay tight. Do not apply glue to the long sides of boards.

Install the first row of flooring with the edge of the tongues directly over the chalk line. Make sure end joints are tight, then wipe up any excess glue immediately. At walls, leave a ½" space to allow for expansion of the wood. This gap will be covered by the baseboard and base shoe.

For succeeding rows, insert the tongue into the groove of the preceding row, and pivot the flooring down into the adhesive. Gently slide the tongue and groove ends together. At walls, use a hammer and a flooring pull bar to draw together the joints on the last strip (inset).

After you've installed three or four rows, use a mallet and scrap piece of flooring to gently tap boards together, closing up the seams. All joints should fit tightly.

Use a cardboard template to fit boards in irregular areas. Cut cardboard to match the space, and allow for a ½" expansion gap next to the wall. Trace the template outline on a board, then cut it to fit using a jig saw. Finish layering strips over the entire floor.

Bond the flooring to the adhesive by rolling it with a heavy floor roller. Roll the flooring within 3 hours of the adhesive application. Work in sections, and finish by installing the flooring in the section between your starting line and the wall.

Installing a Decorative Medallion

If anything is more beautiful under your feet than a newly installed hardwood floor, it's a decorative centerpiece that complements the rest of the surface. Ready-made hardwood medallions, such as the one shown in this project, are relatively easy to install, and provide a focal point for the entire room.

How to Install a Decorative Medallion

Place the medallion on the floor where you want it installed. Draw a line around the medallion onto the floor.

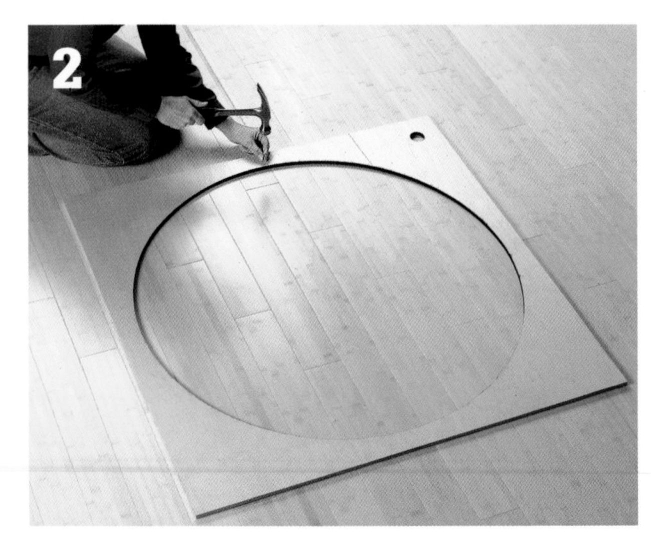

Nail the installation jig to the floor so the opening is aligned with the outline you drew in the previous step. Drive the nails into joints in the floor.

Using the router bit that came with the medallion, place the bearing of the router bit on the inside edge of the jig opening and make a ¼"-deep cut. Remove any exposed nails or staples. Make repeated passes with the router, gradually increasing the depth.

Use a pry bar to remove the flooring inside the hole. Remove all nails. Dry-fit the medallion to ensure it fits. Remove the jig and fill nail holes with wood putty.

Apply urethane flooring adhesive to the subfloor where the hardwood was removed. Spread the adhesive with a trowel. Set the medallion in place and push it firmly into the adhesive so it's level with the surrounding floor.

Laminate

Laminate flooring comes in a floating system that is simple to install, even if you have no experience with other home-improvement projects. You may install a floating laminate floor right on top of plywood, concrete slab, sheet vinyl, or hardwood flooring. Just be sure to follow the manufacturer's instructions.

The pieces are available in planks or squares in a variety of different sizes, colors, and faux finishes—including wood and ceramic. The part you see is really a photographic print. Tongue-and-groove edges lock pieces together, and the entire floor floats on the underlayment. At the end of this project there are a few extra steps to take if your flooring manufacturer recommends using glue on the joints.

The rich wood tones of beautiful laminate planks may cause you to imagine hours of long, hard installation work, but this is a DIY project that you can do in a single weekend. Buy the manufactured planks at a home-improvement or flooring store and install laminate flooring with the step-by-step instructions offered in the following pages.

Tools & Materials ▸

- Circular saw
- Underlayment
- ½" spacers
- Tapping block
- Scrap foam
- Speed square
- Manufacturer glue
- Painter's tape
- Chisel
- Rubber mallet
- Drawbar
- Finish nails
- Nail set strap clamps
- Threshold and screws

How to Install a Floating Floor

To install the underlayment, start in one corner and unroll the underlayment to the opposite wall. Cut the underlayment to fit, using a utility knife or scissors. Overlap the second underlayment sheet according to the manufacturer's recommendations, and secure the pieces in place with adhesive tape.

Working from the left corner of the room to right, set wall spacers and dry lay planks (tongue side facing the wall) against the wall. The spacers allow for expansion. If you are flooring a room more than 26 ft. long or wide, you need to buy appropriate-sized expansion joints. *Note: Some manufacturers suggest facing the groove side to the wall.*

Set a new plank right side up, on top of the previously laid plank, flush with the spacer against the wall at the end run. Line up a speed square with the bottom plank edge and trace a line. That's the cutline for the final plank in the row.

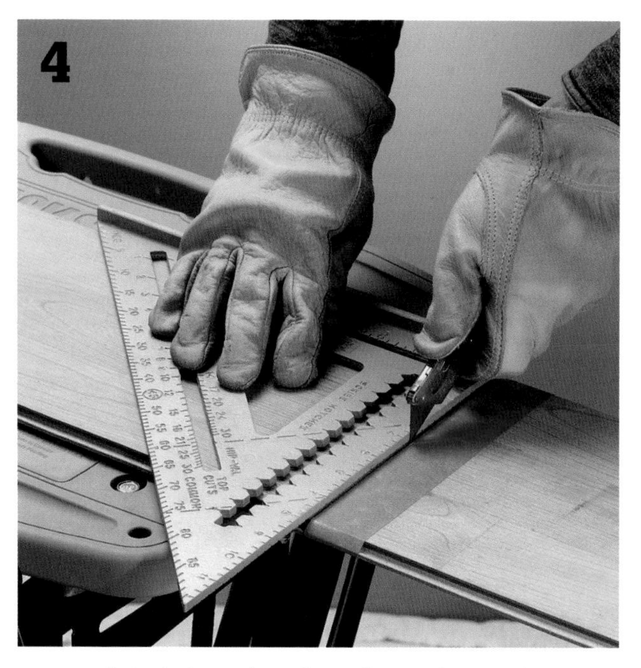

Press painter's tape along the cutline on the top of the plank to prevent chips when cutting. Score the line drawn in Step 3 with a utility knife. Turn the plank over and extend the pencil line to the backside.

(continued)

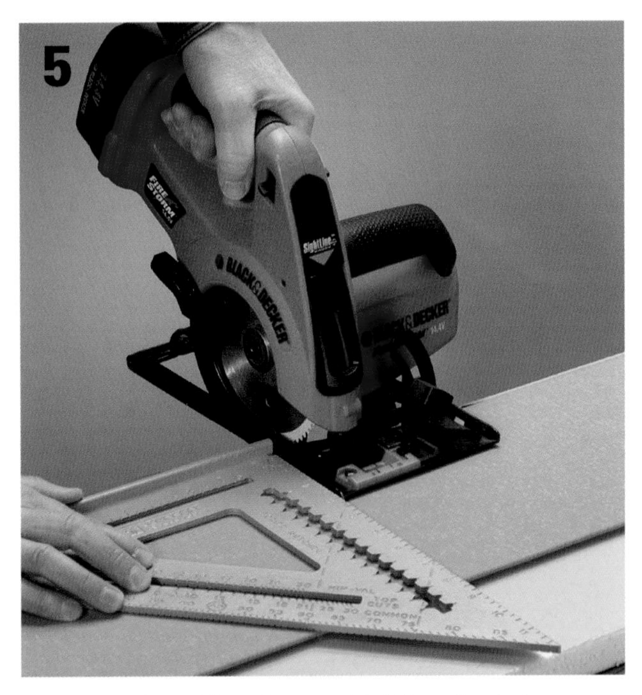

Clamp the board face down and rigid foam insulation or plywood to a work table. The foam reduces chipping. Clamp a speed square on top of the plank, as though you are going to draw another line parallel to the cutline—use this to eye your straight cut. Place the circular saw's blade on the waste side of the actual cutline.

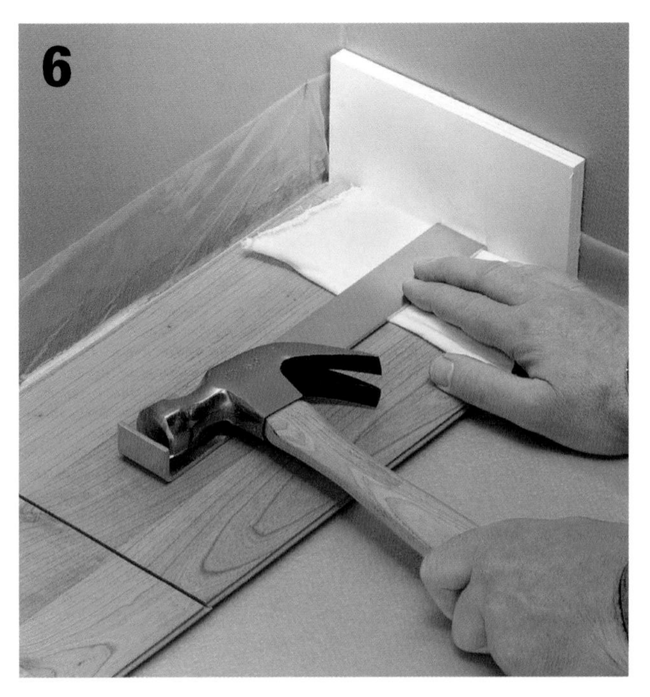

To create a tight fit for the last plank in the first row, place a spacer against the wall and wedge one end of a drawbar between it and the last plank. Tap the other end of the drawbar with a rubber mallet or hammer. Protect the laminate surface with a thin cloth.

Continue to lay rows of flooring, making sure the joints are staggered. This prevents the entire floor from relying on just a few joints, which keeps the planks from lifting. Staggering also stengthens the floor, because the joints are shorter and more evenly distributed.

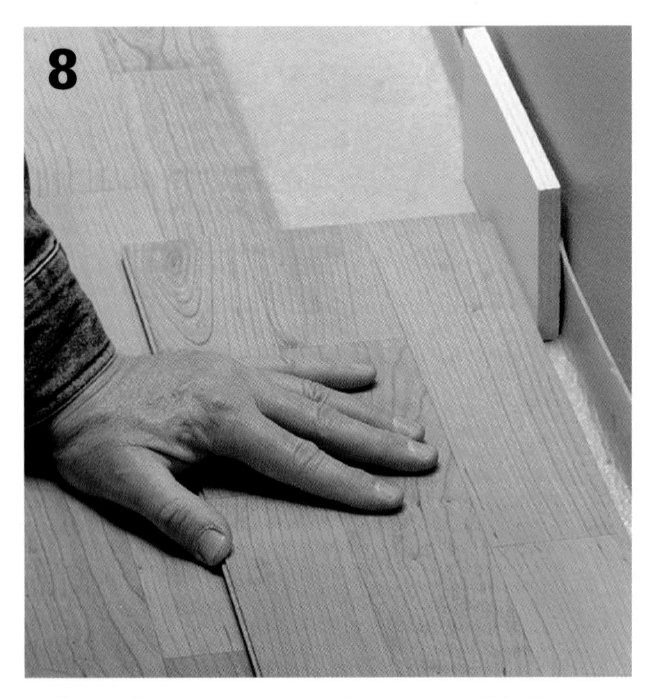

To fit the final row, place two planks on top of the last course; slide the top plank up against the wall spacer. Use the top plank to draw a cutline lengthwise on the middle plank. Cut the middle plank to size using the same method as in Step 3, just across the grain. The very last board must be cut lengthwise and widthwise to fit.

How to Work Around Obstacles

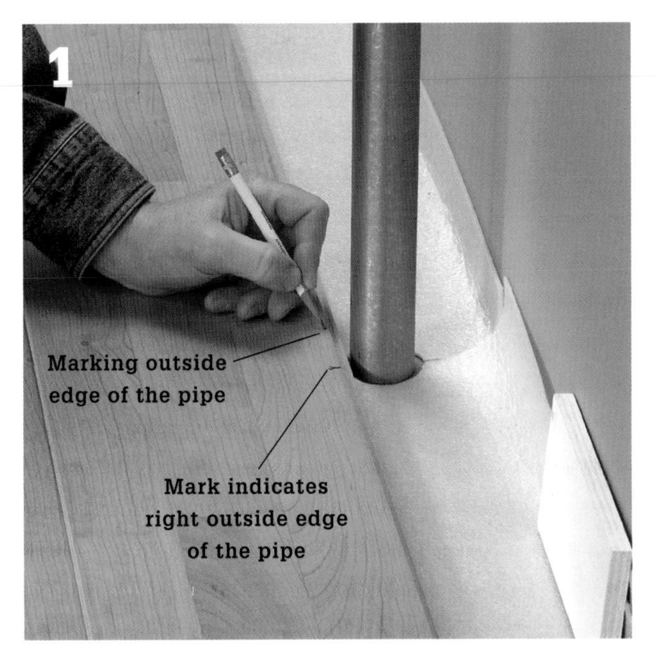

Position a plank end against the spacers on the wall next to the obstacle. Use a pencil to make two marks along the length of the plank, indicating the points where the obstacle begins and ends.

Once the plank is snapped into the previous row, position the plank end against the obstacle. Make two marks with a pencil, this time on the end of the plank to indicate where the obstacle falls along the width of the board.

Use a speed square to extend the four lines. The space at which they intersect is the part of the plank that needs to be removed to make room for the obstacle to go through it. Use a drill with a Forstner bit, or a hole saw the same diameter as the space within the intersecting lines, and drill through the plank at the X. You'll be left with a hole; extend the cut to the edges with a jig saw.

Install the plank by locking the tongue-and-groove joints with the preceding board. Fit the end piece in behind the pipe or obstacle. Apply manufacturer-recommended glue to the cut edges, and press the end piece tightly against the adjacent plank. Wipe away excess glue with a damp cloth.

How to Install Laminate Flooring Using Adhesive

Dry-fit each row, then completely fill the groove of the plank with the glue supplied or recommended by the manufacturer.

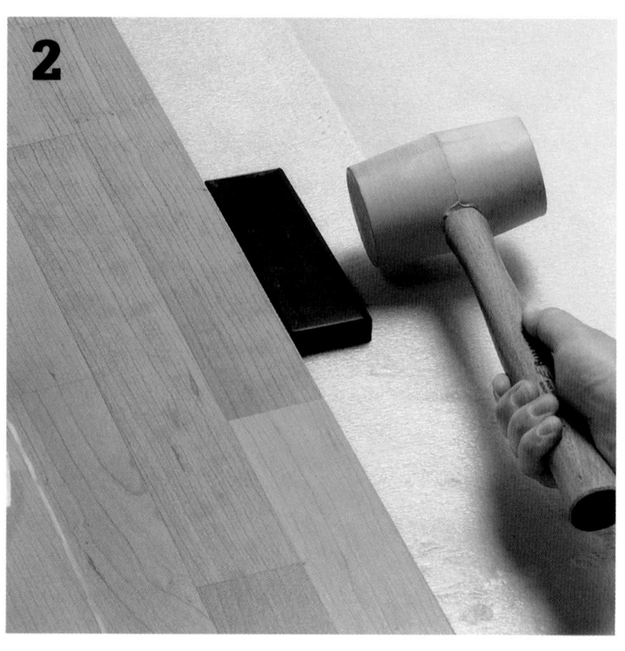

Close the gaps between end joints and lengthwise joints, using a rubber mallet and block to gently tap the edge or end of the last plank. Use a drawbar for the last planks butted up to a wall. Wipe away excess glue in the joints with a damp cloth before it dries.

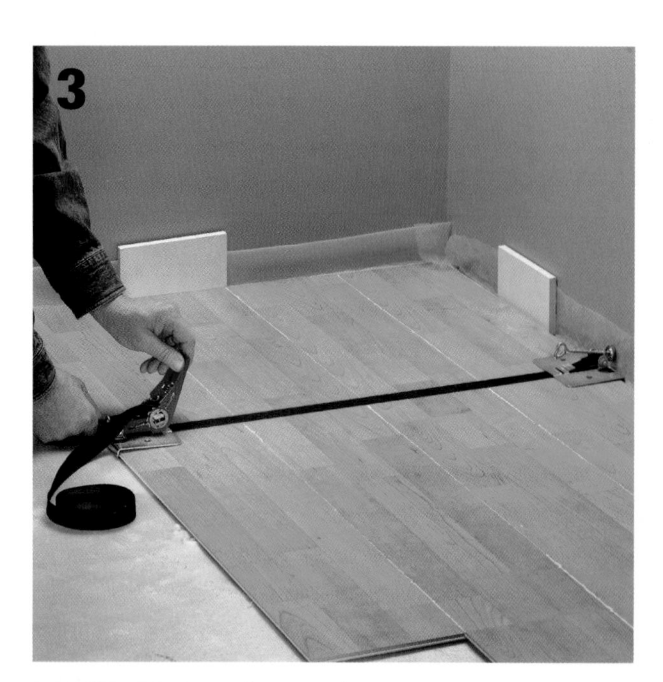

Rent 6 to 10 strap clamps to hold a few rows of planks together as adhesive dries (about an hour). Fit one end of the strap clamp over the plank nearest the wall, and the other end (the one with the ratchet lever) over the last plank. Use the ratchet to tighten straps until joints are snug.

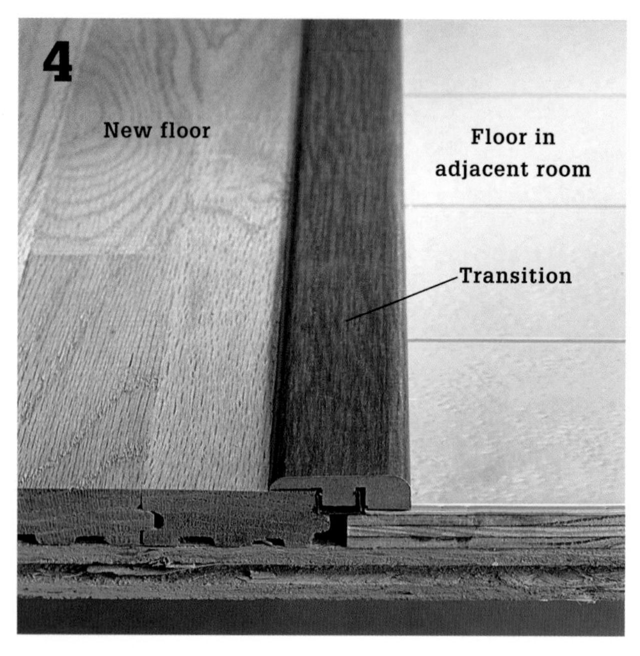

Install transition thresholds at room borders where the new floor joins another floor covering. These thresholds are used to tie together dissimilar floor coverings, such as laminate floorings and wood or carpet. They may also be necessary to span a distance in height between flooring in one room and the next.

Installation Tips ▸

Ripcut planks from the back side to avoid splintering the top surface. For accurate straight cuts, mark the cut with a chalk line. If your pencil line is not straight, double-check your tracing—your wall may not be not perfectly straight, in which case you should cut along your hand-drawn pencil line.

Place another piece of flooring next to the piece marked for cutting to provide a stable surface for the foot of the saw. Also, clamp a cutting guide to the planks at the correct distance from the cutting line to ensure a straight cut.

If you need to cut a plank to fit snugly against another plank or a wall with an obstacle in the middle, such as a heat vent, measure in to the appropriate cutline to fit the board flush with the adjacent board or wall (on the other side of the obstacle). Draw a line across the plank in this location. Then measure the obstacle and transfer those measurements to the plank. Drill a starter hole just large enough to fit your jig saw blade into it. Cut the plank along the drawn lines, using a jig saw. Set the board in place by locking the tongue-and-groove joints with the preceding board.

Parquet

For a hardwood floor with greater design appeal, consider installing a parquet floor. It offers more visual interest than strip flooring without sacrificing the beauty and elegance of wood. Parquet comes in a variety of patterns and styles to create geometric designs. It can range from elaborate, custom-designed patterns on the high end, to the more common herringbone pattern, to the widely available and less expensive block design.

Parquet has experienced a radical transformation over the years. A few years ago, each individual piece of parquet was hand-cut and painstakingly assembled piece by piece. Today, parquet is prefabricated so the individual pieces making up the design are available as single tiles, which not only has reduced the cost, but has made the flooring easier to install.

Many types and designs of parquet floors are available, from custom-made originals to standard patterns, but they are all installed the same way—set in adhesive on a wood subfloor. The effort can be very rewarding: Parquet can be used to create shapes and decorations not possible with other wood flooring.

The finger block pattern is one of the most widely available parquet coverings and also one of the least expensive. The configuration of perpendicular strips of wood emphasizes the different grains and natural color variations.

Tools & Materials ▸

- Tape measure
- Chalk line
- Carpenter's square
- Parquet flooring
- Adhesive
- Notched trowel
- Putty knife
- Rubber mallet
- 100- to 150-pound floor roller
- Jig saw
- Solvent

How to Install Parquet Flooring

Mark the centerpoint of each wall. Snap chalk lines between the marks on opposite walls to establish your reference lines. Use the 3-4-5 triangle method to check the lines for squareness (see page 83).

Lay out a dry run of panels from the center point along the reference lines to adjacent walls. If more than half of the last panel needs to be cut off, adjust the lines by half the width of the panel. Snap new working lines, if necessary.

Put enough adhesive on the subfloor for your first panel, using a putty knife. Spread the adhesive into a thin layer with a notched trowel held at a 45° angle. Apply the adhesive right up to the working lines, but do not cover them.

Place the first panel on the adhesive so two sides are flush with the working lines. Take care not to slide or twist the panel when setting it into place. This panel must be positioned correctly to keep the rest of your floor square.

(continued)

Apply enough adhesive for six to eight panels and spread it with a notched trowel.

Set the next panel in place by holding it at a 45° angle and locking the tongue-and-groove joints with the first panel. Lower the panel onto the adhesive without sliding it. Install remaining panels the same way.

After every six to eight panels are installed, tap them into the adhesive with a rubber mallet.

For the last row, align panels over the top of the last installed row. Place a third row over the top of these, with the sides butted against ½" spacers along the wall. Draw a line along the edge of the third panels onto the second row, cut the panels at the marks, and install.

To work around corners or obstacles, align a panel over the last installed panel, then place another panel on top of it as in step 8. Keep the top panel ½" from the wall or obstacle and trace along the opposite edge onto the second panel (top). Move the top two panels to the adjoining side, making sure not to turn the top panel. Make a second mark on the panel the same way (bottom). Cut the tile with a jig saw and install.

Within 4 hours of installing the floor, roll the floor with a 100- to 150-pound floor roller. Wait at least 24 hours before walking on the floor again.

How to Install Parquet with a Diagonal Layout

Establish perpendicular working lines following Step 1 on page 104. Measure 5 ft. from the centerpoint along each working line and make a mark. Snap chalk lines between the 5 ft. marks. Mark the centerpoint of these lines, then snap a chalk line through the marks to create a diagonal reference line.

Lay out a dry run of tiles along a diagonal line. Adjust your starting point as necessary. Lay the flooring along the diagonal line using adhesive, following the steps for installing parquet (pages 104 to 106). Make paper templates for tile along walls and in corners. Transfer the template measurements to tiles, and cut to fit.

Resilient Sheet Vinyl

Preparing a perfect underlayment is the most important phase of resilient sheet vinyl installation. Cutting the material to fit the contours of the room is a close second. The best way to ensure accurate cuts is to make a cutting template. Some manufacturers offer template kits, or you can make one by following the instructions on the opposite page. Be sure to use the recommended adhesive for the sheet vinyl you are installing. Many manufacturers require that you use their glue for installation. Use extreme care when handling the sheet vinyl, especially felt-backed products, to avoid creasing and tearing.

Tools & Materials ▸

- Linoleum knife
- Framing square
- Compass
- Scissors
- Non-permanent felt-tipped pen
- Utility knife
- Straightedge
- 1/4" V-notched trowel
- J-roller
- Stapler
- Flooring roller
- Chalk line
- Heat gun
- 1/16" V-notched trowel
- Straightedge
- Vinyl flooring
- Masking tape
- Heavy butcher or brown wrapping paper
- Duct tape
- Flooring adhesive
- 3/8" staples
- Metal threshold bars
- Nails

Tools for Resilient Floors

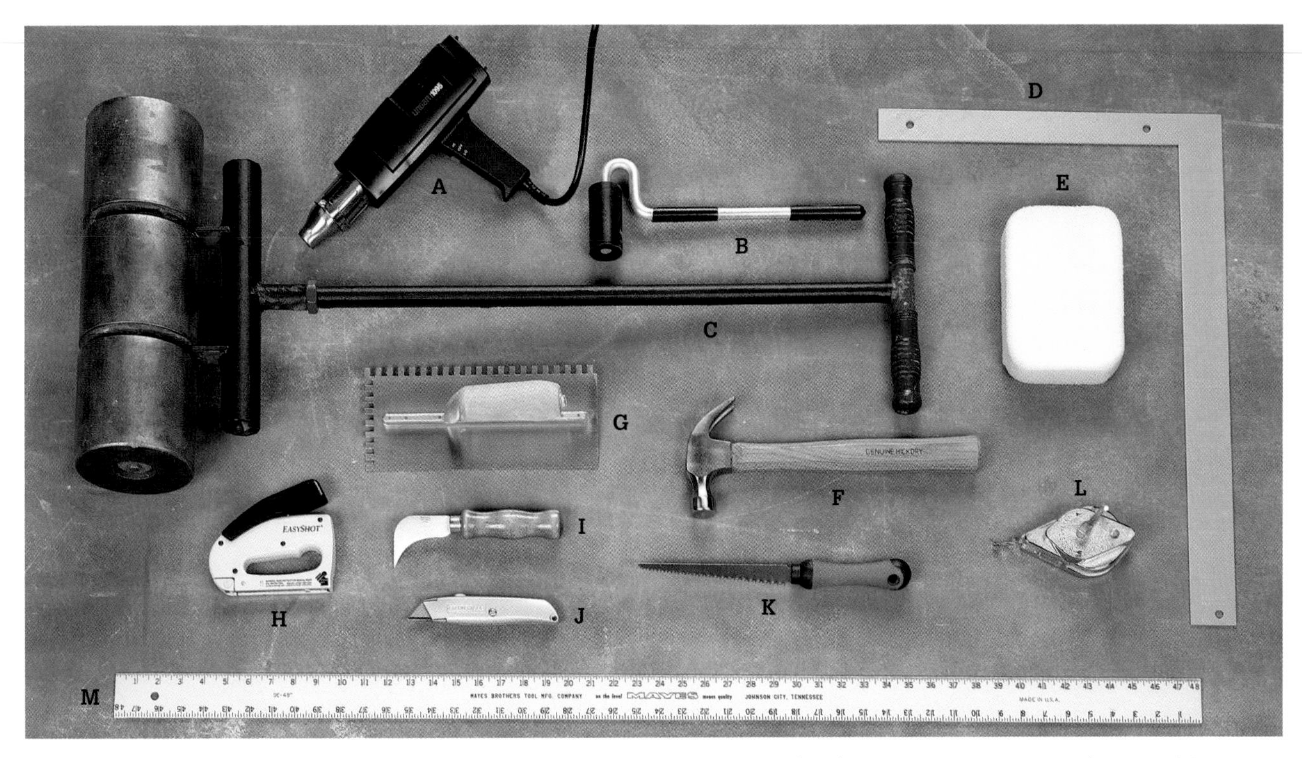

Tools for resilient flooring include: a heat gun (A), J-roller (B), floor roller (C), framing square (D), sponge (E), hammer (F), notched trowel (G), stapler (H), linoleum knife (I), utility knife (J), wallboard knife (K), chalk line (L), straightedge (M).

Buying & Estimating

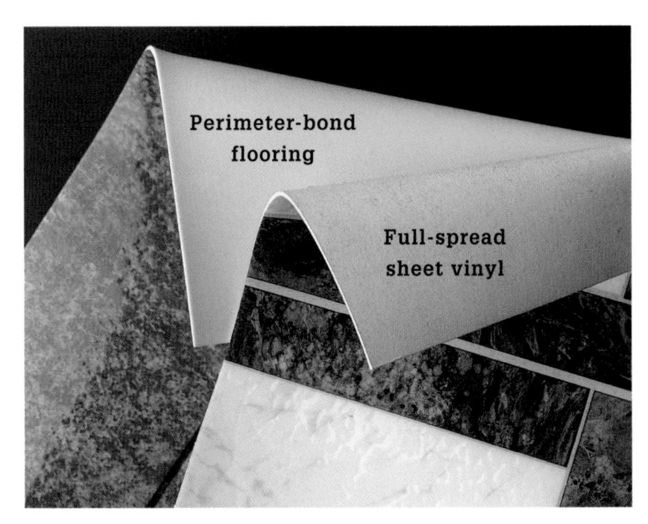

Resilient sheet vinyl comes in full-spread and perimeter-bond styles. Full-spread sheet vinyl has a felt-paper backing and is secured with adhesive that is spread over the floor before installation. Perimeter-bond flooring, identifiable by its smooth, white PVC backing, is laid directly on underlayment and is secured by a special adhesive spread along the edges and seams.

Resilient tile comes in self-adhesive and dry-back styles. Self-adhesive tile has a pre-applied adhesive protected by wax paper backing that is peeled off as the tiles are installed. Dry-back tile is secured with adhesive spread onto the underlayment before installation. Self-adhesive tile is easier to install than dry-back tile, but the bond is less reliable. Don't use additional adhesives with self-adhesive tile.

Installation Tips ▸

Sweep and vacuum the underlayment thoroughly before installing resilient flooring to ensure a smooth, flawless finish (left). Small pieces of debris can create noticeable bumps in the flooring (right).

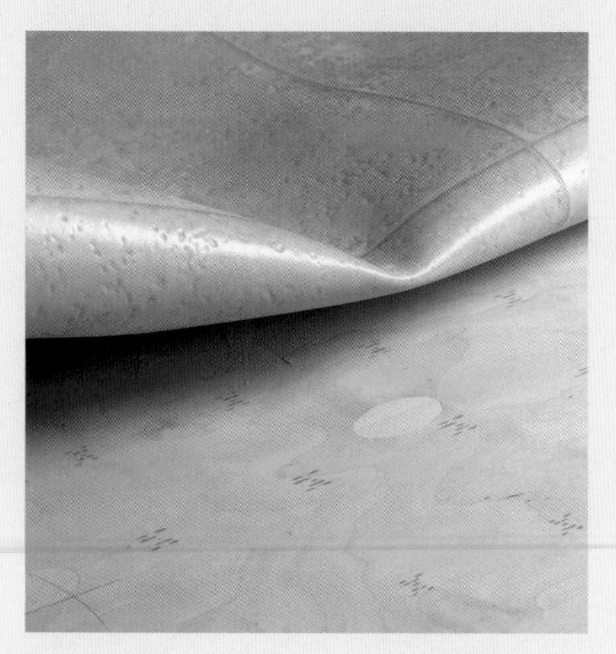

Handle resilient sheet vinyl carefully to avoid creasing or tearing. Working with a helper can help prevent costly mistakes. Make sure the sheet vinyl is at room temperature before you handle it.

How to Make a Cutting Template

Place sheets of heavy butcher paper or brown wrapping paper along the walls, leaving a ⅛" gap. Cut triangular holes in the paper with a utility knife. Fasten the template to the floor by placing masking tape over the holes.

Follow the outline of the room, working with one sheet of paper at a time. Overlap the edges of adjoining sheets by about 2" and tape the sheets together.

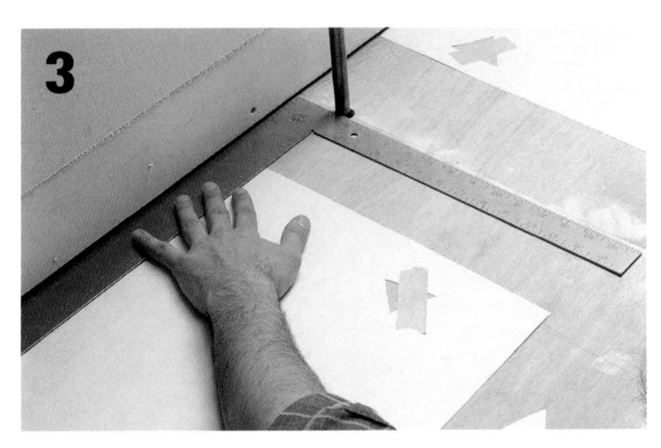

To fit the template around pipes, tape sheets of paper on either side. Measure the distance from the wall to the center of the pipe, then subtract ⅛".

Transfer the measurement to a separate piece of paper. Use a compass to draw the pipe diameter on the paper, then cut out the hole with scissors or a utility knife. Cut a slit from the edge of the paper to the hole.

Fit the hole cutout around the pipe. Tape the hole template to the adjoining sheets.

When completed, roll or loosely fold the paper template for carrying.

How to Install Perimeter-bond Sheet Vinyl

Unroll the flooring on any large, flat, clean surface. To prevent wrinkles, sheet vinyl comes from the manufacturer rolled with the pattern-side out. Unroll the sheet and turn it pattern-side up for marking.

For two-piece installations, overlap the edges of the sheets by at least 2". Plan to have the seams fall along the pattern lines or simulated grout joints. Align the sheets so the pattern matches, then tape the sheets together with duct tape.

Position the paper template over the sheet vinyl and tape it in place. Trace the outline of the template onto the flooring using a non-permanent felt-tipped pen.

Remove the template. Cut the sheet vinyl with a sharp linoleum knife or a utility knife with a new blade. Use a straightedge as a guide for making longer cuts.

Cut holes for pipes and other permanent obstructions. Cut a slit from each hole to the nearest edge of the flooring. Whenever possible, make slits along pattern lines.

Roll up the flooring loosely and transfer it to the installation area. Do not fold the flooring. Unroll and position the sheet vinyl carefully. Slide the edges beneath undercut door casings.

Cut the seams for two-piece installations using a straightedge as a guide. Hold the straightedge tightly against the flooring, and cut along the pattern lines through both pieces of vinyl flooring.

Remove both pieces of scrap flooring. The pattern should now run continuously across the adjoining sheets of flooring.

(continued)

Fold back the edges of both sheets. Apply a 3" band of multipurpose flooring adhesive to the underlayment or old flooring, using a ¼" V-notched trowel or wallboard knife.

Lay the seam edges one at a time onto the adhesive. Make sure the seam is tight, pressing the gaps together with your fingers, if needed. Roll the seam edges with a J-roller or wallpaper seam roller.

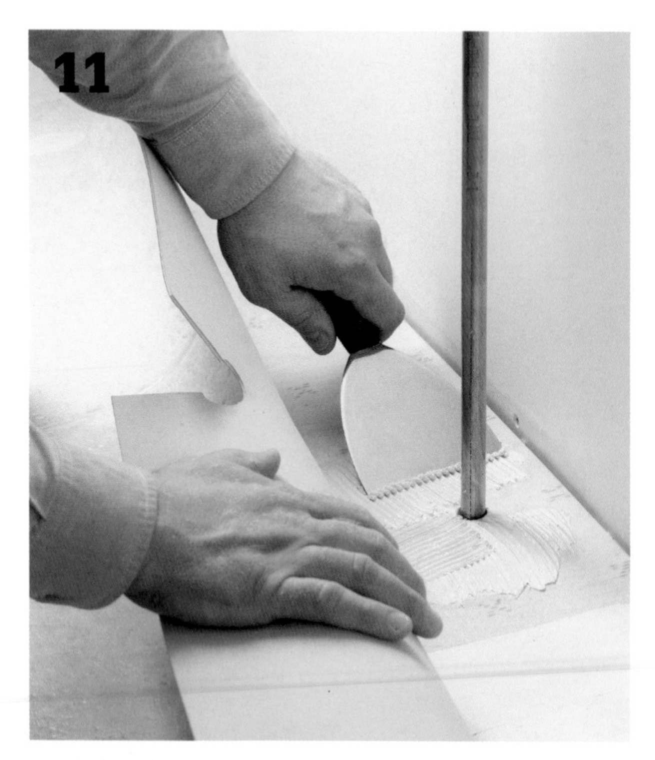

Apply flooring adhesive underneath flooring cuts at pipes or posts and around the entire perimeter of the room. Roll the flooring with the roller to ensure good contact with the adhesive.

If you're applying flooring over a wood underlayment, fasten the outer edges of the sheet with ⅜" staples driven every 3". Make sure the staples will be covered by the base molding.

How to Install Full-spread Sheet Vinyl

Cut the sheet vinyl using the techniques described on pages 130 and 131 (steps 1 to 5), then lay the sheet vinyl into position, sliding the edges under door casings.

Pull back half of the flooring, then apply a layer of flooring adhesive over the underlayment or old flooring using a ¼" V-notched trowel. Lay the flooring back onto the adhesive.

The roller creates a stronger bond and eliminates air bubbles. Fold over the unbonded section of flooring, apply adhesive, then replace and roll the flooring. Wipe up any adhesive that oozes up around the edges of the vinyl, using a damp rag.

Measure and cut metal threshold bars to fit across doorways. Position each bar over the edge of the vinyl flooring and nail it in place.

Rubber Roll

Once a mark of restaurants and retailers, sheet rubber flooring has become an option for homeowners as well. It's resilient, durable, and stable, holding up well under the heaviest and most demanding use. Better still, it's comfortable to walk on and easy to maintain.

The durability and resilience of rubber provide benefits in two ways. First, the flooring takes just about any kind of use without showing damage. Second, it absorbs shock in proportion to its thickness. Heavier rubber floors help prevent fatigue, making them comfortable for standing, walking, and even strenuous exercise.

Many new flooring products are made from recycled rubber, which saves landfill space and reduces the consumption of new raw materials. This is one place a petroleum-based product is environmentally friendly.

To install rubber sheet flooring on top of wood, use only exterior-grade plywood, one side sanded. Do not use lauan plywood, particleboard, chipboard, or hardboard. Make sure the surface is level, smooth, and securely fastened to the subfloor.

Tools & Materials ▸

- Adhesive
- Chalk line
- Cleaning supplies
- Craft/utility knife
- Flat-edged trowel
- Measuring tape
- Mineral spirits
- Notched trowel
- Painter's tape
- Straightedge
- Weighted roller

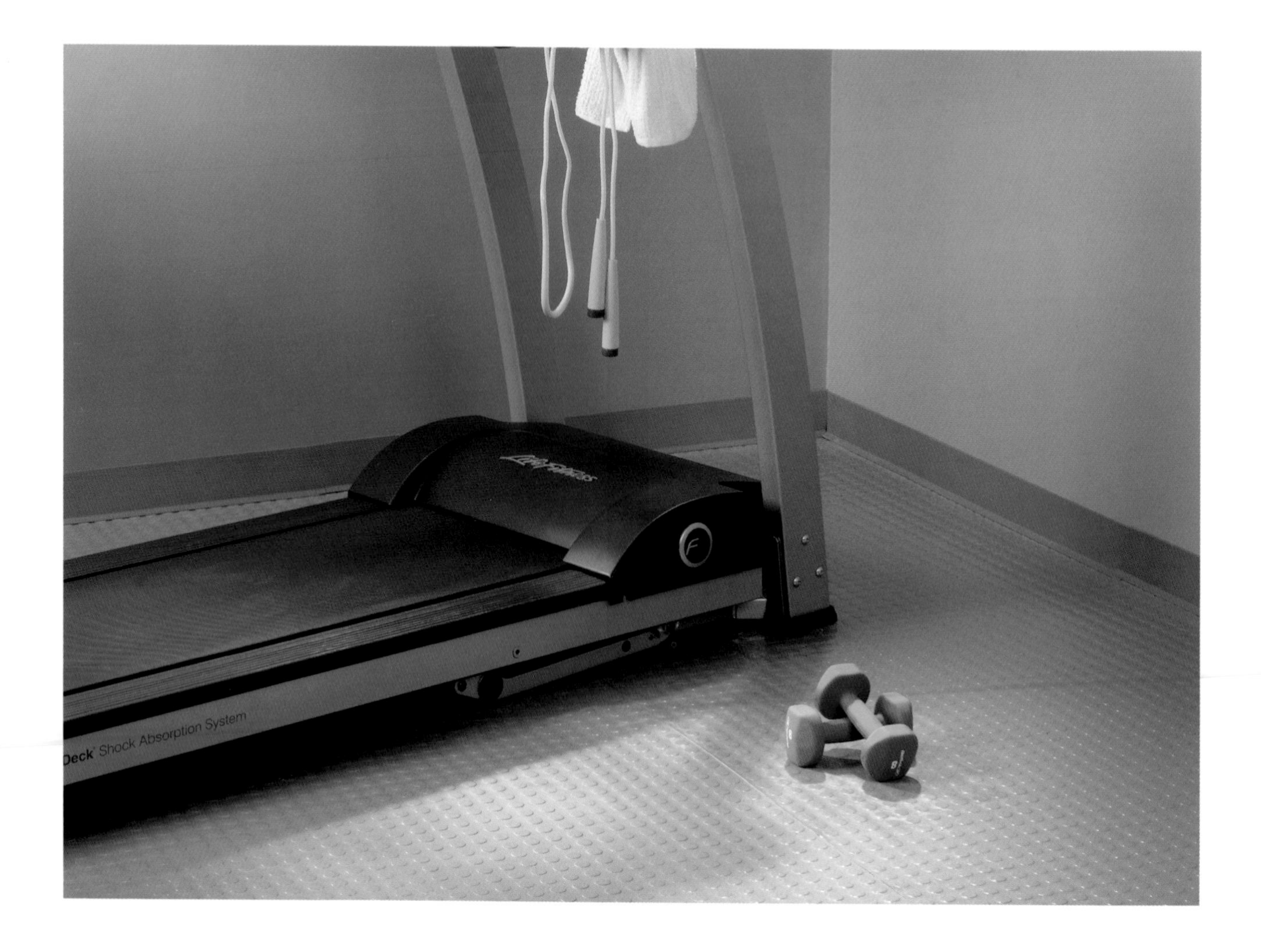

How to Install a Rubber Roll

Measure the longest wall in the room; this is where you will place the first roll. Cut the roll to length, leaving a few extra inches at each end. This lets you cut the ends accurately once the material has relaxed.

Position the first sheet against the longest wall, and square it with the room. Cut the second length of material, and position it to overlap with the first roll by 1 to 1½ inches at the seams. Repeat this process until the rolls cover the room from end to end. *Note: Allow the cut rolls to relax in position for at least 2 hours before you begin seaming and gluing. Then use one of the following methods for creating clean, straight seams.*

For materials that are 4mm to 6mm thick, place a 4-inch wide scrap of material under the seam area. Using a straight edge and a craft knife with a new razor blade, hold the knife straight up and down and slice through both pieces in one cut. *Tip: For materials that are 9mm thick or more, snap a chalk line on the first roll where the seam should be. Align the leading edge of the second roll with the chalk line. Lift only the edge of the second roll.*

Using a straightedge and craft knife, cut along the chalk line. The cut must be perfectly straight. Then lower the second roll into place and butt it against the first roll. Do not try to compress or stretch the material. Use these methods to complete all the seams. Once you are finished, the rolls should be ready to fasten in place. Many adhesives are available, but not all are compatible. Use only the type of adhesive recommended by the manufacturer.

(continued)

Lift the first roll lengthwise, halfway back from the wall. Using the method recommended by the manufacturer, apply adhesive to the underlayment and spread it with a notched trowel or other approved tool.

When you apply adhesive, use only as much as you can apply and roll at one time. The average working time is about 30 minutes at 70°F and 50% relative humidity. Trowels have limited useful lives, too. Use a new one for each container of adhesive.

Lower the roll slowly onto the adhesive, making sure not to allow any air to become trapped underneath. Never leave adhesive ridges or puddles; they will become visible on the surface.

Roll the floor immediately with a 100-pound roller to squeeze out any trapped air and maximize contact between the roll and the adhesive. With each pass of the roller, overlap the previous pass by half. Roll the width first, then the length, and re-roll after 30 minutes.

Fold back the second half of the first roll, and the first half of the second roll. Apply and spread the adhesive as before. Spread the adhesive at a 90-degree angle to the seams. This will reduce the chance of having adhesive squeeze up through the seams.

Continue the process, rolling each section as soon as the material is set in the adhesive. If you get adhesive on the top surface, clean it off quickly with dilute mineral spirits.

Hand-roll all seams after the entire floor has been rolled. If you see gaps in the seams, hold them together temporarily with painter's tape. For the thinnest materials, you may want to weigh down the seams until they are fully set. Long boxes of trimwork are useful for this job.

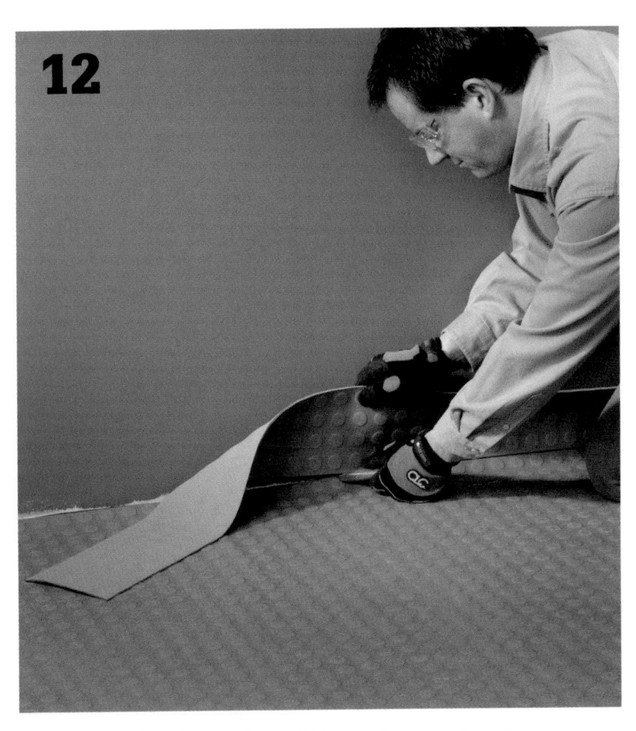

Give the adhesive at least 24 hours to cure, then lay a straight edge along the uncut ends of the rolls and trim them carefully with a craft knife. The very ends should be covered by baseboard and base shoe.

Resilient Tile

As with any tile installation, resilient tile requires carefully positioned layout lines. Before committing to any layout and applying tile, conduct a dry run to identify potential problems.

Keep in mind the difference between reference lines (see opposite page) and layout lines. Reference lines mark the center of the room and divide it into quadrants. If the tiles don't lay out symmetrically along these lines, you'll need to adjust them slightly, creating layout lines. Once layout lines are established, installing the tile is a fairly quick process. Be sure to keep joints between the tiles tight and lay the tiles square.

Tiles with an obvious grain pattern can be laid so the grain of each tile is oriented identically throughout the installation. You can also use the quarter-turn method, in which each tile has its pattern grain running perpendicular to that of adjacent tiles. Whichever method you choose, be sure to be consistent throughout the project.

Tools & Materials ▸

- Tape measure
- Chalk line
- Framing square
- Utility knife
- 1/16" notched trowel
- Heat gun
- Resilient tile
- Flooring adhesive (for dry-back tile)

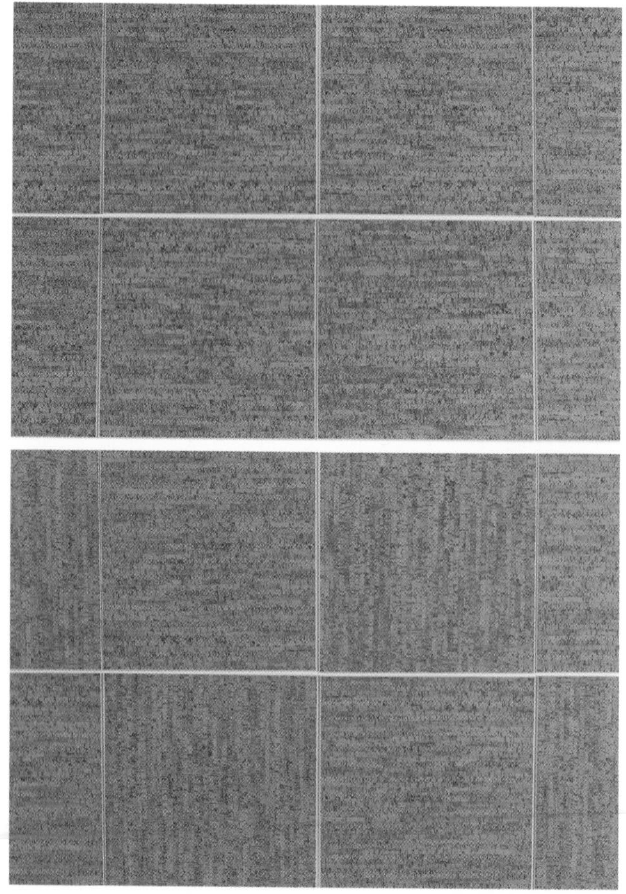

Check for noticeable directional features, like the grain of the vinyl particles. You can set the tiles in a running pattern so the directional feature runs in the same direction (top), or in a checkerboard pattern using the quarter-turn method (bottom).

How to Make Reference Lines for Tile Installation

Position a reference line (X) by measuring along opposite sides of the room and marking the center of each side. Snap a chalk line between these marks.

Measure and mark the centerpoint of the chalk line. From this point, use a framing square to establish a second reference line perpendicular to the first one. Snap the second line (Y) across the room.

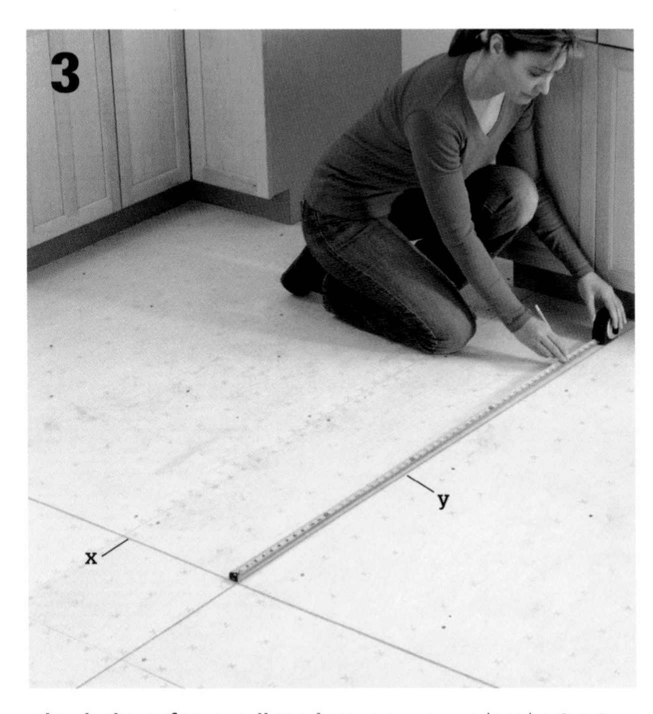

Check the reference lines for squareness using the 3-4-5 triangle method. Measure along reference line X and make a mark 3 ft. from the centerpoint. Measure from the centerpoint along reference line Y and make a mark at 4 ft.

Measure the distance between the marks. If the reference lines are perpendicular, the distance will measure exactly 5 ft. If not, adjust the reference lines until they're exactly perpendicular to each other.

How to Install Dry-backed Resilient Tile

Snap perpendicular reference lines with a chalk line. Dry-fit tiles along layout line Y so a joint falls along reference line X. If necessary, shift the layout to make the layout symmetrical or to reduce the number of tiles that need to be cut.

If you shift the tile layout, create a new line that is parallel to reference line X and runs through a tile joint near line X. The new line, X1, is the line you'll use when installing the tile. Use a different colored chalk to distinguish between lines.

Dry-fit tiles along the new line, X1. If necessary, adjust the layout line as in steps 1 and 2.

If you adjusted the layout along X1, measure and make a new layout line, Y1, that's parallel to reference line Y and runs through a tile joint. Y1 will form the second layout line you'll use during installation.

Apply adhesive around the intersection of the layout lines using a trowel with 1/16" V-shaped notches. Hold the trowel at a 45° angle and spread adhesive evenly over the surface.

Spread adhesive over most of the installation area, covering three quadrants. Allow the adhesive to set according to the manufacturer's instructions, then begin to install the tile at the intersection of the layout lines. You can kneel on installed tiles to lay additional tiles.

When the first three quadrants are completely tiled, spread adhesive over the remaining quadrant, then finish setting the tile.

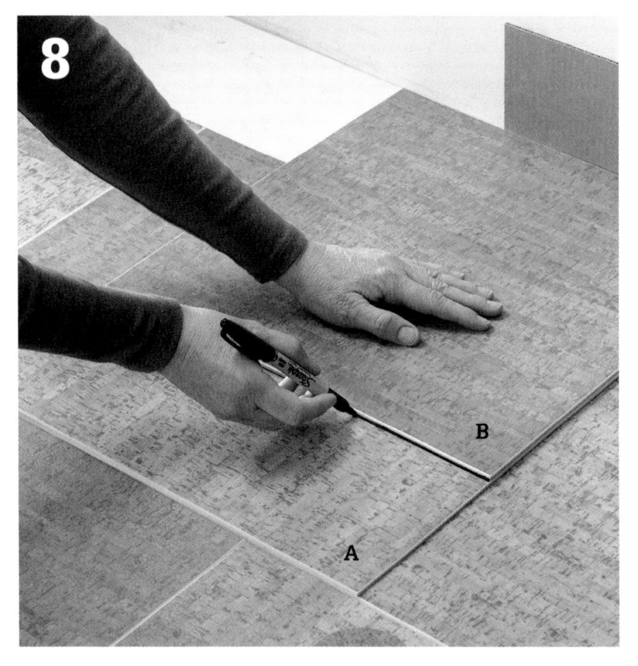

To cut tiles to fit along the walls, place the tile to be cut (A) face up on top of the last full tile you installed. Position a 1/8"-thick spacer against the wall, then set a marker tile (B) on top of the tile to be cut. Trace along the edge of the marker tile to draw a cutting line.

(continued)

Tip ▸

To mark tiles for cutting around outside corners, make a cardboard template to match the space, keeping a ⅛" gap along the walls. After cutting the template, check to make sure it fits. Place the template on a tile and trace its outline.

9

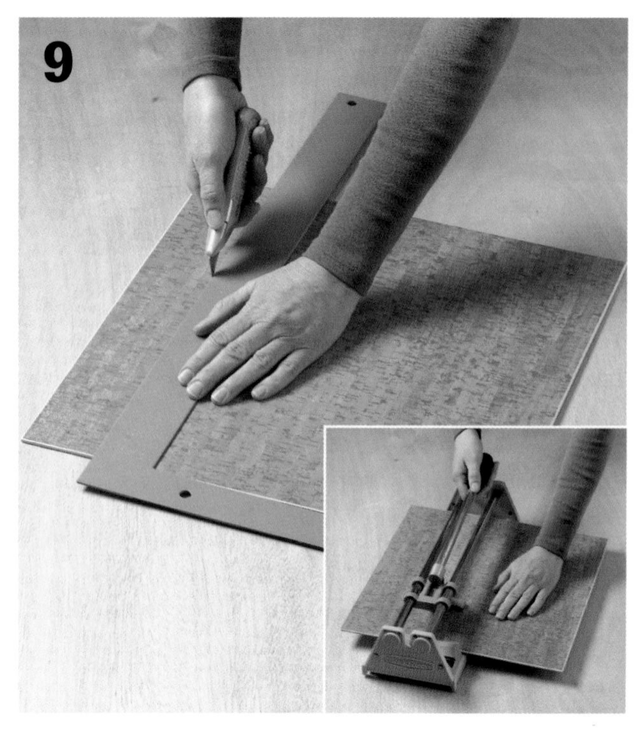

Cut tile to fit using a utility knife and straightedge. Hold the straightedge securely against the cutting line to ensure a straight cut. *Option: You can use a ceramic-tile cutter to make straight cuts in thick vinyl tiles (see inset).*

10

Install cut tiles next to the walls. If you're precutting all tiles before installing them, measure the distance between the wall and install tiles at various points in case the distance changes.

11

Continue installing tile in the remaining quadrants until the room is completely covered. Check the entire floor. If you find loose areas, press down on the tiles to bond them to the underlayment. Install metal threshold bars at room borders where the new floor joins another floor covering.

How to Install Self-adhesive Resilient Tile

Once your reference lines are established, peel off the paper backing and install the first tile in one of the corners formed by the intersecting layout lines. Lay three or more tiles along each layout lines in the quadrant. Rub the entire surface of each tile to bond the adhesive to the floor underlayment.

Begin installing tiles in the interior area of the quadrant. Keep the joints tight between tiles..

Finish setting full tiles in the first quadrant, then set the full tiles in an adjacent quadrant. Set the tiles along the layout lines first, then fill in the interior tiles.

Continue installing the tile in the remaining quadrants until the room is completely covered. Check the entire floor. If you find loose areas, press down on the tiles to bond them to the underlayment. Install metal threshold bars at room border where the new floor joins another floor covering.

Combination Tile

This hybrid product combines the classic, refined look of ceramic tile with the easy installation of resilient flooring. Made to resemble a range of hard materials, from slate to quarry tile to marble, combination tiles feel warmer and more comfortable underfoot than ordinary ceramic tile.

Designs vary from brand to brand, but most major manufacturers of resilient tile now offer these combination products. You can install them like regular resilient tile, each placed tightly against the next, or you can leave spaces between the squares and add grout—giving it even more of the look and feel of regular ceramic tile.

With or without grout, combination tile is easy to maintain. Some manufacturers offer generous warranties, promising that the tiles will not fade, stain, crack, or wear through for many years. This confidence translates to peace of mind for you. Consider these products if you want to look beyond traditional choices.

Tools & Materials ▸

- Tape measure
- Chalk line
- Framing square
- Utility knife
- 1⁄16" notched trowel
- Combination tile
- Flooring adhesive
- Weighted roller
- Joint sealer

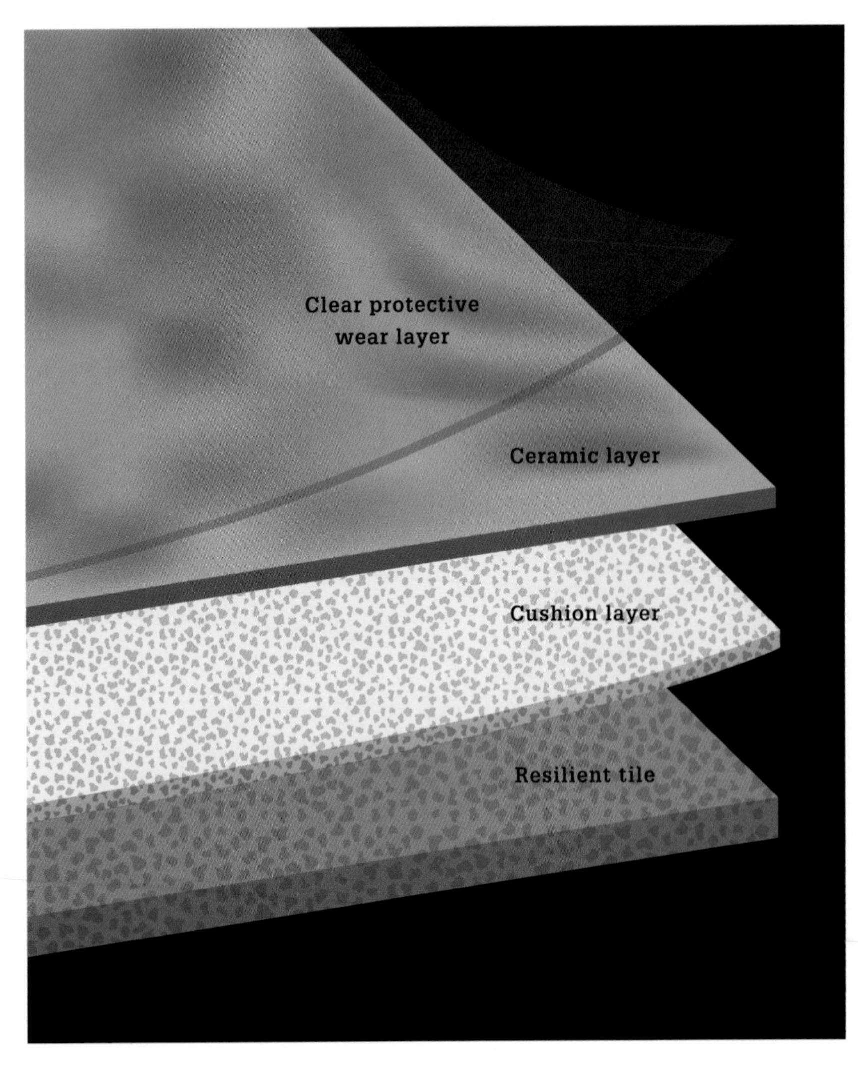

In kitchens and bathrooms, floor coverings need to withstand heavy traffic, frequent cleaning, and lots of moisture. Ceramic tile meets these needs, but it can also be difficult to install, it's cold underfoot, and unforgiving with dropped dishes. Vinyl, in sheets or tiles, makes a softer, warmer surface that is inexpensive and relatively easy to install. But vinyl is vulnerable to scrapes and gouges, and doesn't last as long as ceramic tile. In recent years, manufacturers have found ways of using the best properties of both materials in combination tiles. These are vinyl tiles covered with a thin layer of ceramic composite. They can be installed like regular vinyl tiles, with their edges pushed together, or with spaces left for grout.

How to Install Combination Tile

Compound resilient tiles can be installed on a variety of surfaces. Check the manufacturer's instructions to make sure your underlayment is recommended. It should be clean, dry, and free of dust, dirt, grease, and wax. Sweep, vacuum, and damp-mop the surface before you begin.

Measure the outside edges of the room, and snap chalk lines across the center. Dry-lay a row of tiles from the center to each wall (be sure to incorporate spacers). If the last row will be less than one-fourth the width of one tile, adjust the center point to balance the layout.

Find the middle point on opposite walls, and snap chalk lines between them. The intersection of these lines should be in the middle of the room. Check the intersecting lines for square, using the 3-4-5 method.

Starting at the intersection, dry-lay a row of tiles to one wall. If the last tile will be less than one-quarter the width of a full tile, you may want to move the center point.

Avoid positioning tile joints directly over underlayment joints or seams in existing flooring. If this happens, reposition the chalk lines to offset joints by at least 3" or half the width of one tile. Repeat the dry-laying test, adjusting the lines as needed, until you have a definite starting point.

Apply the recommended adhesive to one quadrant of the center intersection and spread it with a notched trowel or other approved tool. Let it stand for the time specified by the manufacturer. Use only as much adhesive as you can cover during the working time allowed. Continue to work from the center outward in each quadrant.

To work around obstacles, place the tile up against the obstacle and mark for cut lines. Follow manufacturer instructions for cutting tile.

Within an hour after the tiles are set, roll the floor with a weighted roller. Work in both directions, taking care not to push any tiles out of place. Re-roll the floor before grouting the tiles or applying a joint sealer.

Bamboo Planks

It looks like hardwood, and is available in traditional tongue-and-groove form and in laminate planks. But bamboo is not wood. It's really a grass—and one of the most popular flooring materials today.

Bamboo flooring is made by shredding stalks of the raw material, then pressing them together with a resin that holds the shreds in their finished shape. Not only is bamboo a fast-growing and renewable crop, the companies that make bamboo flooring use binders with low emissions of volatile organic compounds (VOCs). The result is tough, economical, and ecologically friendly. In other words, it's just about perfect for flooring.

If you choose tongue-and-groove bamboo, the installation techniques are the same as for hardwoods. Bamboo is also available as a snap-fit laminate for use in floating floors. In this project we show Teragren Synergy Strand in Java (see Resources): thin, durable planks that are glued to the underlayment.

Tools & Materials ▸

- Adhesive
- Carpenter's level
- Carpenter's square
- Chalk line
- Cleaning supplies
- Flat-edged trowel
- Marking pen or pencil
- Measuring tape
- Moisture level meter
- Notched trowel
- Rubber mallet
- Scrap lumber
- Shims
- Straightedge
- Weighted roller

Tips for a Successful Installation ▸

Bamboo plank flooring should be one of the last items installed on any new construction or remodeling project. All work involving water or moisture should be completed before floor installation. Room temperature and humidity of installation area should be consistent with normal, year-round living conditions for at least a week before installation. Room temperature of 60 to 70°F and humidity range of 40 to 60% is recommended.

About Radiant heat: The subfloor should never exceed 85°F. Check the manufacturer's suggested guidelines for correct water temperature inside heating pipes. Switch on the heating unit three days before flooring installation. Room temperature should not vary more than 15°F year-round. For glue-down installations, leave the heating unit on for three days following installation.

How to Install Bamboo Planks

Give the bamboo time to adjust to installation conditions. Store it for at least 72 hours in or near the room where it will be installed. Open the packages for inspection, but do not store the planks on concrete or near outside walls.

Even though thin-plank bamboo is an engineered material, it can vary in appearance. Buy all planks from the same lot and batch number. Then visually inspect the planks to make sure they match. Use the same lighting as you will have in the finished room.

Inspect the underlayment. Bamboo planks can be installed on plywood or oriented strand board at least ¾ inch thick. The underlayment must be structurally sound; wood surfaces should have no more than 12 percent moisture.

Make sure the underlayment is level. It should not change by more than ⅛ inch over 10 feet. If necessary, apply a floor leveler to fill any low places, and sand down any high spots. Prevent squeaks by driving screws every 6 inches into the subfloor below.

Sweep, vacuum, and damp-mop the surface, then measure all room dimensions. If the longest facing walls are parallel, begin installing the planks on one side of the room. For irregular shapes and uneven walls, establish a straight starting line next to one long wall and work from there.

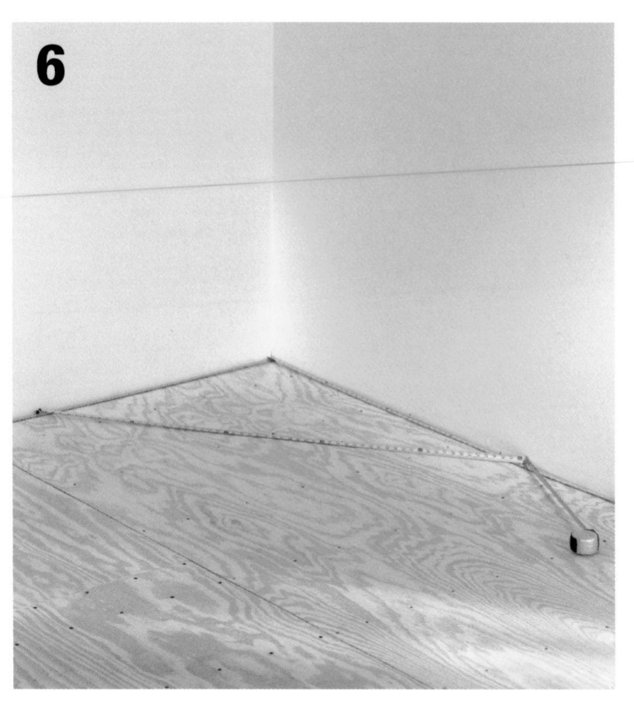

The planks should be perpendicular to the floor joists below. Adjust your starting point if necessary. Snap a chalk line next to the longest wall. The distance from the wall should be the same at both ends, leaving ½ inch for expansion.

Lay the first course of planks with the tongue edge toward the wall. Align the planks with the chalk line. Hold the edge course in place with wedges, or by nailing through the tongue edge. This row will anchor the others, so make sure it stays securely in place.

Once the starter row is in place, install the planks using a premium wood flooring adhesive. Be sure to follow the manufacturer's instructions. Begin at the chalk line and apply enough adhesive to lay down one or two rows of planks. Spread the adhesive with a V-notched trowel at a 45-degree angle. Let the adhesive sit for the specified time.

(continued)

When the adhesive is tacky and ready to use, lay the first section of bamboo planks. Set each plank in the adhesive by placing a clean piece of scrap lumber on top and tapping it down with a rubber mallet. Check the edge of each section to make sure it keeps a straight line.

After you finish the first section, cover the next area with adhesive and give it time to become tacky. This slows down the project, but it prevents you from using more adhesive than you can use—and it allows the section you just finished to set up.

When the adhesive is ready, lay down the next section of planks. Fit the new planks tightly against the previous section, taking care not to knock the finished section out of alignment. If the planks have tongue-and-groove edges, fit them carefully into place.

Continue applying adhesive and installing planks, one section at a time, to cover the entire floor. When adhesive gets on the flooring surface, wipe it off quickly.

At the edges and around any fixed objects, such as doorways or plumbing pipes, leave a ½-inch gap for expansion. Use shims to maintain the gaps if needed. These spaces can be covered with baseboards, base shoe, and escutcheons.

As you finish each section, walk across it a few times to maximize contact between the planks and the adhesive. When all the planks are in place, clean the surface and use a clean weighted roller. Push the roller in several directions, covering the entire surface many times.

In places that are difficult to reach with a roller, such as the edges of the room, lay down a sheet of protective material, such as butcher paper, and stack weights on the paper to press down on the planks.

Let the finished floor sit for at least 24 hours, then clean the surface carefully and remove any spacers from the expansion gaps. Finally, install the finishing trim.

Cork Tile

Cork flooring has all kinds of benefits. It dampens sound, resists static, insulates surfaces, and provides visual warmth like no other floor covering. Cork is a renewable natural product—tree bark—that can be harvested once a decade without cutting down the tree.

Natural cork tiles complement most furnishings and decorations, and can be found in every shade from honey yellow to deep espresso brown. Left in its original color or stained, it has beautiful patterns that range from burls to spalting. Every tile is different, which means no two installations will look the same.

The tiles may seem fragile, but they take on the strength of the underlayment below. In fact, once a cork floor is installed and properly sealed, it can withstand normal household use just as well as any other material.

Tools & Materials ▸

- Tape measure
- Chalk line
- Framing square
- Utility knife
- Notched trowel
- Cork tile
- Primer
- Recommended flooring adhesive
- Scrap block
- Rubber mallet
- Paint roller and tray
- Joint sealer

Tips for a Successful Installation ▸

Before you work with any cork flooring material, remove it from the package and leave it in the room where it will be installed. This lets the material adjust to the room's temperature and humidity. Manufacturers recommend acclimating cork for at least 72 hours.

How to Install Cork Tile

If you plan to install the cork on plywood underlayment or a similar material, make sure the surface is clean and dry, with no more change in level than ⅛ inch over 10 feet. Fill any low spots and sand down any high spots.

Measure the outside edges of the room, and snap chalk lines across the center. Dry-lay a row of tiles from the center to each wall.

If the last row will be less than one-fourth the width of one tile, adjust the center point to balance the layout.

Apply a recommended adhesive, using the method specified by the manufacturer. Some adhesives are best applied with a paint roller, others with a notched trowel. Put down only as much adhesive as you can use in the time allowed.

Cork adhesive needs to air out for 20 to 30 minutes before you can begin laying the cork tiles on it. After that, the working time is roughly one hour. Check the time allowed on the adhesive you choose.

Begin at the intersection of the two center guidelines, and work on one quadrant at a time. Apply enough adhesive to hold as many tiles as you can place in the working time allowed. Set the first tile in place, and check to make sure the adhesive holds it firmly.

(continued)

Cork tile colors and patterns will vary. Use this to your advantage by mixing batches for greater variety.

If the humidity is high during installation, fit the tiles together tightly so they won't pull apart in drier weather. To fit each new tile in place, hold a piece of scrap lumber against the edge and tap it gently with a rubber mallet.

Work from the center outward in each quadrant. As you complete each row, check to see that all edges are straight. If a row has gone out of line, remove as many tiles as necessary and start again. It's frustrating to make corrections now, but it's difficult or impossible later.

To fit tile around an object, such as a plumbing pipe, cut a tile-sized piece of paper. Work the paper into the space, cutting as needed until it fits. Then lay the paper on a tile and use it as a cutting guide. Patch wide cuts with scrap material.

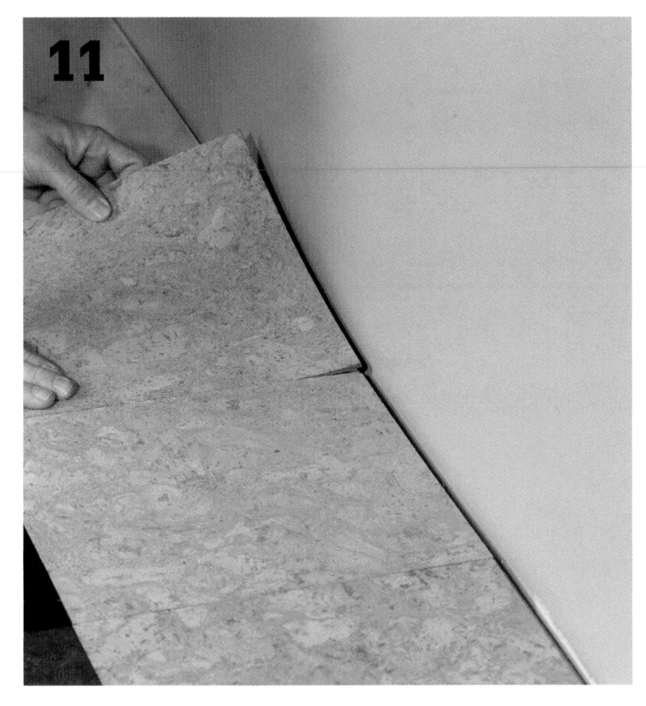

Like other natural materials, cork will expand and contract as the temperature and humidity change. To allow room for varying conditions, leave a ¼-inch gap between the finished floor and all walls, thresholds, water pipes, and other vertical surfaces.

If your cork floor measures more than 30 feet in any direction, install an expansion joint—either in a continuous surface or where the cork meets another flooring material. This allows the cork to flex with changing conditions.

Once all the tiles are in place, vacuum the surface to remove all dust. Make several passes across the entire surface with a 100-pound floor roller. Move the roller in different directions to press every tile down securely. Let the finished floor sit overnight, then roll it again.

After rolling the surface, clean it once more and apply a recommended sealer. The most common types are water-based polyurethane. Make sure the sealer gets in all joints so it prevents moisture from entering and damaging the finished floor. Allow the sealed floor to sit for another day before you use it. Then you can install molding, trim, and other finishing pieces.

Recycled Rubber Tile

You see it in commercial buildings, from office towers to skating rinks. It's recycled rubber made from ground-up tires. Sometimes it's solid black, but most often it has one or two other colors blended in as small flecks. And these days, more people than ever are putting it in their homes.

Used in basements, it takes the chill out of a concrete floor, making laundry rooms and workshops more pleasant places to spend time. In mudrooms and breezeways, it stands up to heavy traffic and messy conditions. It doesn't work quite as well in kitchens, because recycled rubber tends to be porous, which means it soaks up grease and oil. A rubber kitchen floor would quickly become slippery and might develop unpleasant odors.

If you want flooring that resists changing weather conditions, withstands heavy traffic, and helps keep materials out of the waste stream, recycled rubber is an ideal choice.

Tools & Materials ▸

Adhesive
Aviator's snips
Butcher paper for template
Carpenter's square
Chalk line
Cleaning supplies
Craft/utility knife
Flat-edged trowel
Marking pen or pencil
Measuring tape
Mineral spirits
Notched trowel
Painter's tape
Scrap lumber
Shims
Straightedge
Weighted roller
If you plan to use grout between the tiles:
Coarse sponge
Grout sealer
Grouting trowel
Sanded grout
Tile spacers

How to Install Recycled Rubber Tile

If you plan to install rubber tile on plywood underlayment or a similar material, the surface must be level, with no more change than ⅛ inch over 10 feet. Fill any low spots.

Sand down high spots. Make sure the underlayment is clean, smooth, and securely fastened to the subfloor. Check the manufacturer's instructions to be sure the recommended adhesive will work with your underlayment.

The best way to create a consistent appearance is to buy all your tiles from the same lot and batch number. Even then, mix tiles from different boxes to prevent visible changes in the finished result. Many tiles have arrows on the bottom so you can align the patterns on top.

To allow room for changes in temperature and humidity, leave a ¼-inch gap between the finished floor and all walls, thresholds, water pipes, and other vertical surfaces.

(continued)

Measure the length and width of the room to be covered. Find the center of each direction and snap a chalk line across the room. The intersection of these lines is the middle of the room. Check the lines for square, using the 3-4-5 method.

Dry-lay a row of tiles in each direction until you reach all four walls. If the last tile needs to be cut narrower than one-quarter the width of a full tile, adjust the center point and test the layout again.

Starting at one corner of the center intersection, apply the adhesive recommended by the manufacturer and spread it with a notched trowel. Let it sit for the recommended time, usually 30 minutes at 70°F and 50% relative humidity.

Lay the first tile in the corner. Gently twist it into place and press down on it to work out any trapped air—but don't try to stretch or compress it. Check the edges to be sure the tile is square to the guidelines.

Working in one direction at a time, continue to lay tiles in the adhesive. Stop every 30 minutes and roll the tiles with a weighted roller to squeeze out any trapped air and maximize contact between the tile and the adhesive.

Continue laying individual tiles, checking them for square and rolling them at regular intervals. If you get adhesive on the top surface, clean it off quickly with dilute mineral spirits.

When you reach a wall, flip over the next tile. Place it on top of the next-to-last tile and butt it against the wall. Make a mark across the back where it overlaps the previous tile. Move to a work surface and cut the tile along that line.

At a door jamb, place a tile face up where it will go. Lean it against the jamb and mark the point where they meet. Move the tile to find the other cut line, and mark that as well. Flip the tile over, mark the two lines using a carpenter's square, and cut out the corner.

(continued)

To fit tile around an object, such as a plumbing pipe, cut a tile-sized piece of paper. Work the paper into the space, cutting as needed until it fits. Then lay the paper on a tile and use it as a cutting guide. Patch wide cuts with scrap material.

Work from the center outward in each quadrant. As you complete each row, check to see that all edges are straight. Hand-roll all seams after you finish each row. If you see gaps in the seams, hold them together temporarily with painter's tape and then place weight on top.

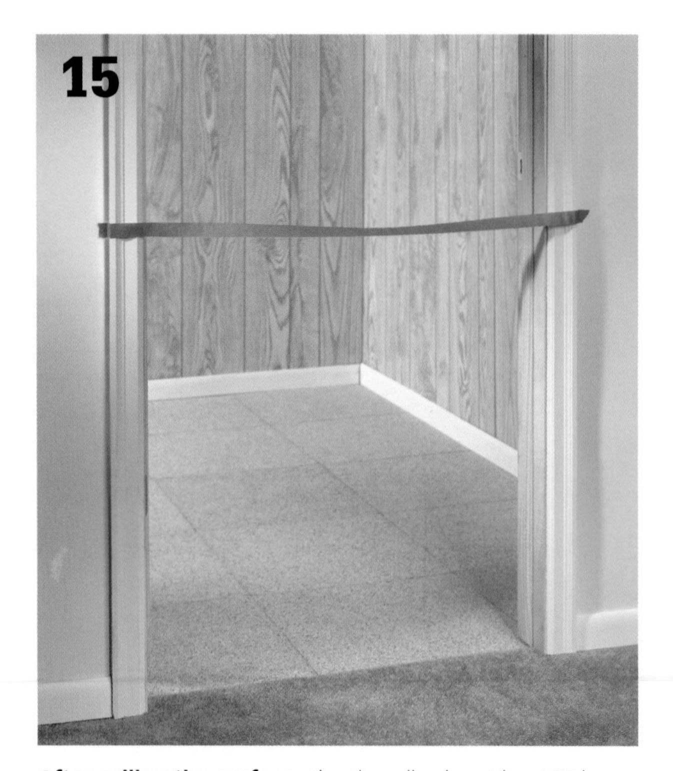

After rolling the surface, give the adhesive at least 24 hours to cure. Foot traffic and rolling loads can cause permanent indentations in the uncured adhesive and cause tiles to shift.

Install any trim pieces, such as baseboards, shoe base, escutcheons, and thresholds.

Ceramic & Stone Tile

Ceramic tile includes a wide variety of hard flooring products made from molded clay. Although there are significant differences between various types, they are all installed using cement-based mortar as an adhesive and grout to fill the gaps between tiles. These same techniques are used to install tiles cut from natural stone, such as granite and marble.

To ensure a long-lasting tile floor, you'll need a smooth, stable, and level subfloor (see page 59). In addition, the underlayment must be solid. Cementboard, or thinner fiber/cementboard, is the best underlayment since it has excellent stability and is unaffected by moisture. Cementboard is manufactured exclusively for ceramic tile installation (see page 60). In rooms where moisture is not a factor, exterior-grade plywood is an adequate underlayment. It's also less expensive. Another option is isolation membrane, which is used to protect ceramic tile and stone from movements caused by cracks in concrete floors. Isolation membrane is used to cover individual cracks, or it can be used to cover an entire floor. Page 61 shows how to install isolation membrane.

Many ceramic tiles have a glazed surface that protects the porous clay from staining. You should protect unglazed ceramic tile from stains and water spots by periodically applying a coat of tile sealer. Keep dirt from getting trapped in grout lines by sealing them once a year.

If you want to install trim tiles, consider their placement as you plan the layout. Some base-trim tile is set on the floor, with its finished edge flush with the field tile. Other types are installed on top of the field tile, after the field tile is laid and grouted.

Ceramic & Stone Tile Tools & Materials

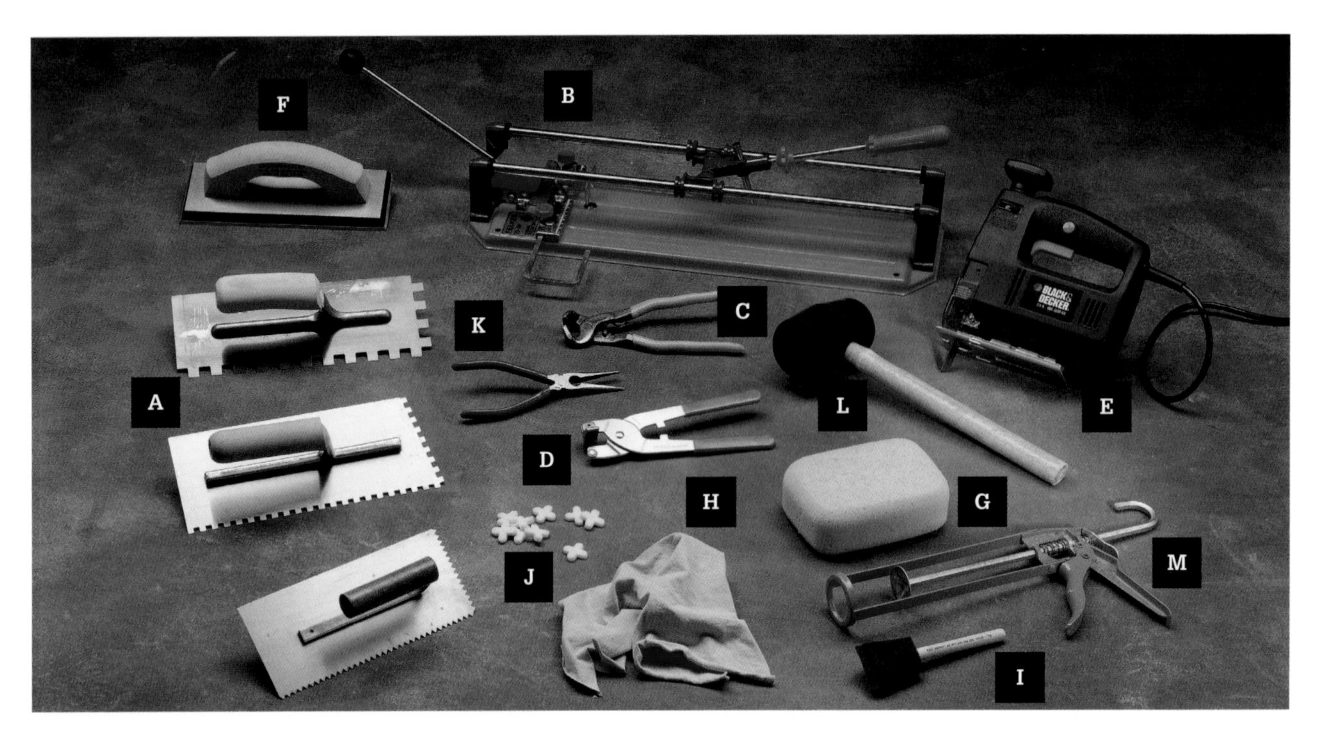

Tile tools include adhesive-spreading tools, cutting tools, and grouting tools. Notched trowels (A) for spreading mortar come with notches of varying sizes and shapes. The size of the notch should be proportional to the size of the tile being installed. Cutting tools include a tile cutter (B), tile nippers (C), hand-held tile cutter (D), and jig saw with carbide blade (E). Grouting tools include a grout float (F), grout sponge (G), buff rag (H), and foam brush (I) for applying grout sealer. Other tile tools include spacers (J), available in different sizes to create grout joints of varying widths; needlenose pliers (K), for removing spacers; rubber mallet (L), for setting tiles into mortar; and caulk gun (M).

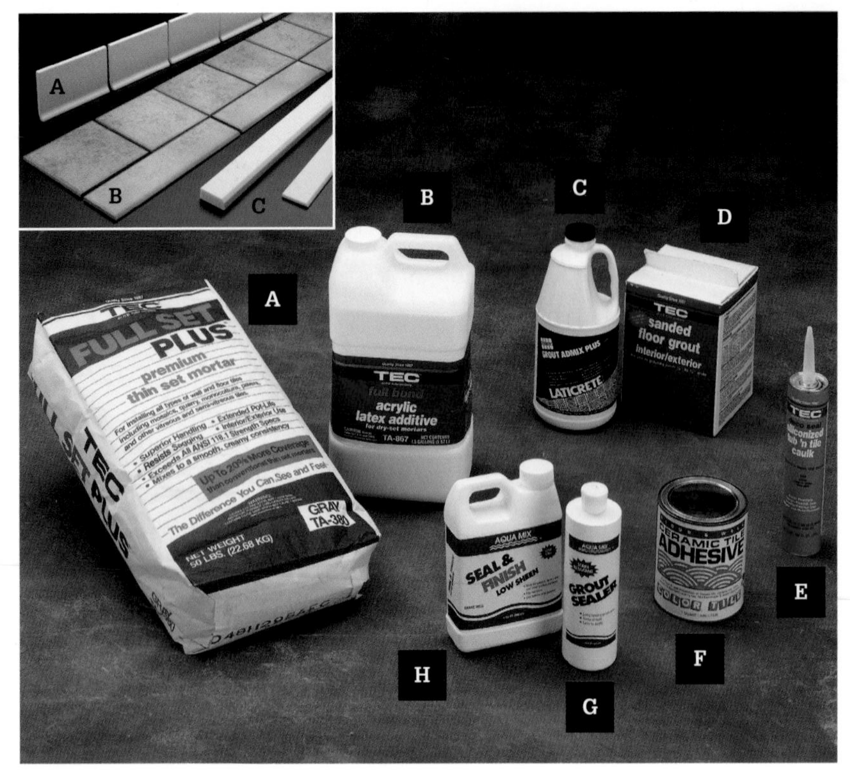

Tile materials include adhesives, grouts, and sealers. Thin-set mortar (A), the most common floor-tile adhesive, is often strengthened with latex mortar additive (B). Grout additive (C) can be added to floor grout (D) to make it more resilient and durable. Grout fills the spaces between tiles and is available in pre-tinted colors to match your tile. Silicone caulk (E) should be used in place of grout where tile meets another surface, like a bathtub. Use wall-tile adhesive (F) for installing base-trim tile. Grout sealer (G) and porous-tile sealer (H) ward off stains and make maintenance easier. (INSET) Trim and finishing materials include base-trim tiles (A), bullnose tiles (B), and doorway thresholds (C) in thicknesses ranging from ¼" to ¾" to match floor levels.

How to Cut Ceramic & Stone Tile

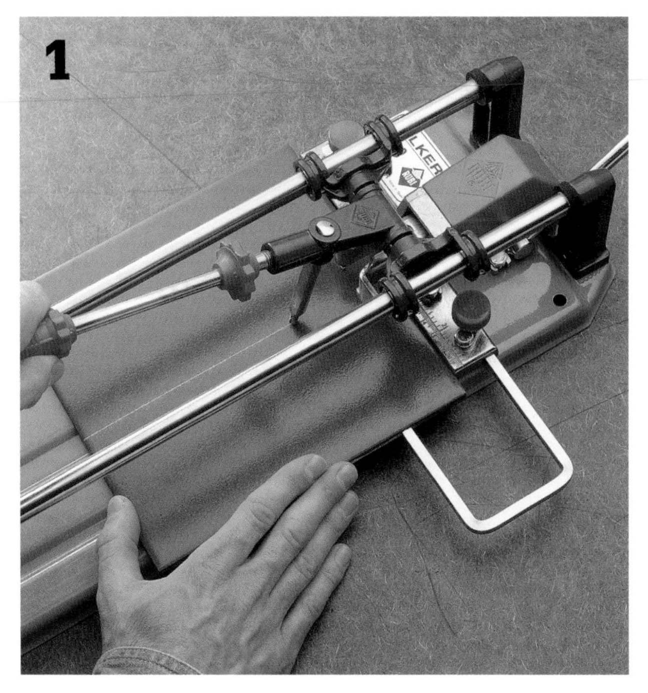

For straight cuts, mark a cutting line on the tile with a pencil, then place the tile in the cutter so the cutting wheel is directly over the line. While pressing down firmly on the wheel handle, run the wheel across the tile to score the surface. For a clean cut, score the tile only once.

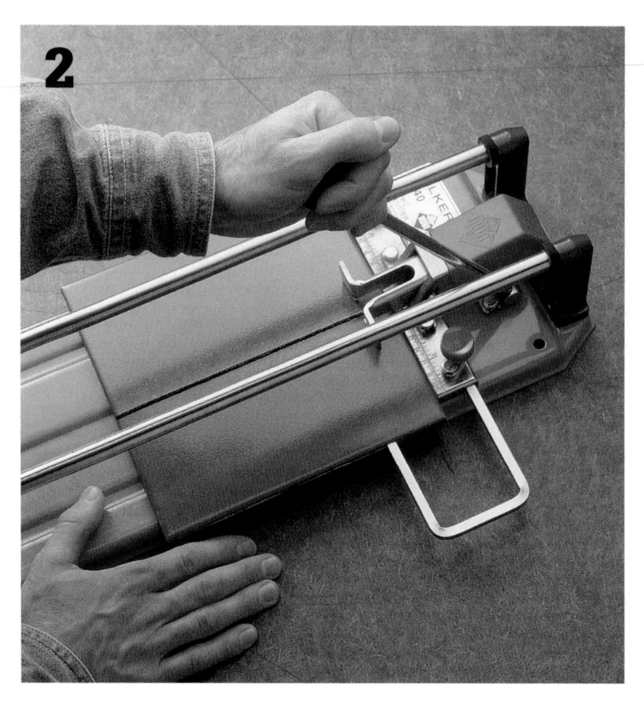

Snap the tile along the scored line as directed by the tool manufacturer. Snapping the tile is usually accomplished by depressing a lever on the tile cutter

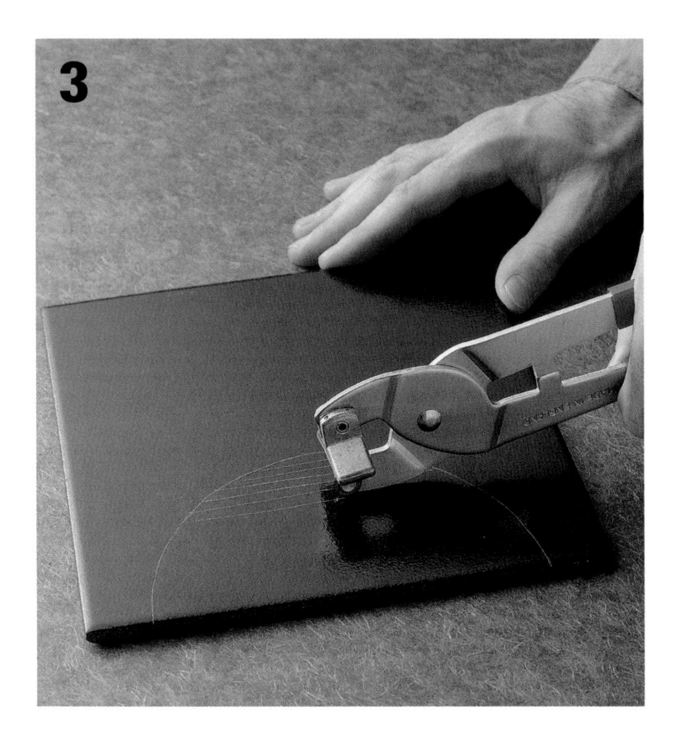

To cut curves, mark a cutting line on the tile face. Use the scoring wheel of a hand-held tile cutter to score the cut line. Make several parallel scores, no more than ¼" apart, in the waste portion of the tile.

Use tile nippers to nibble away the scored portion of the tile. To cut circular holes in the middle of a tile, score and cut the tile so it divides the hole in half, then remove waste material from each half of the circle.

Option: To cut mosaic tiles, use a tile cutter to score tiles in the row where the cut will occur. Cut away excess strips of mosaics from the sheet, using a utility knife, then use a hand-held tile cutter to snap tiles one at a time. Use tile nippers to cut narrow portions of tiles after scoring.

Tip: Using Power Tools for Cutting ▸

Tile saws, also called "wet saws" because they use water to cool blades and tiles, are used primarily for cutting natural-stone tiles. They're also useful for quickly cutting notches in all kinds of hard tile. Wet saws are available for rent at tile dealers and rental shops.

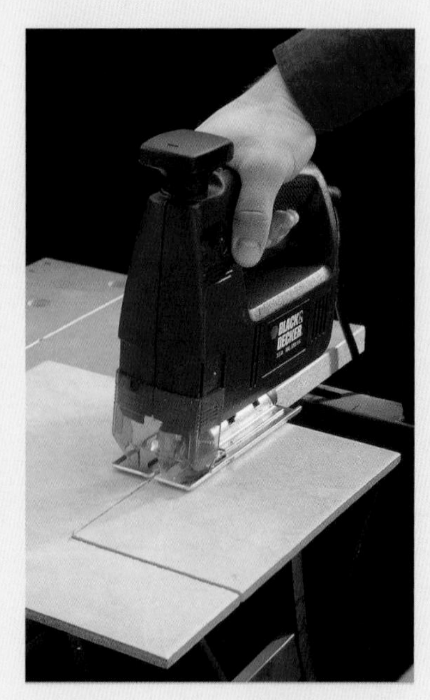

To make square notches, clamp the tile down on a worktable, then use a jig saw with a tungsten-carbide blade to make the cuts. If you need to cut several notches, a wet saw is more efficient.

Cut holes for plumbing stub-outs and other obstructions by marking the outline on the tile, then drilling around the edges using a ceramic tile bit. Gently knock out the waste material with a hammer. The rough edges of the hole will be covered by protective plates on fixtures called escutcheons.

Installing Ceramic Tile

Ceramic tile installation starts with the same steps as installing resilient tile. You snap perpendicular reference lines and dry-fit tiles to ensure the best placement.

When setting tiles, work in small sections so the mortar doesn't dry before the tiles are set. Use spacers between tiles to ensure consistent spacing. Plan an installation sequence to avoid kneeling on set tiles. Be careful not to kneel or walk on tiles until the designated drying period is over.

Tools & Materials ▸

¼" square trowel
Rubber mallet
Tile cutter
Tile nippers
Hand-held tile cutter
Needlenose pliers
Grout float
Grout sponge
Soft cloth
Small paint brush
Thin-set mortar
Tile
Tile spacers
Grout
Latex grout additive
Wall adhesive
2 × 4 lumber
Grout sealer
Tile caulk
Sponge brush

How to Install Ceramic Tile

Make sure the subfloor is smooth, level, and stable. Spread thin-set mortar on the subfloor for one sheet of cementboard. Place the cementboard on the mortar, keeping a ¼" gap along the walls.

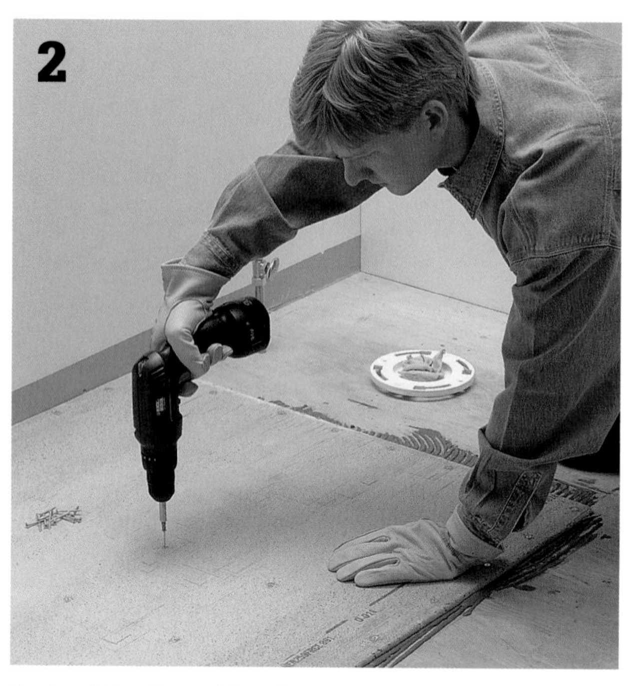

Fasten it in place with 1¼" cementboard screws. Place fiberglass-mesh wallboard tape over the seams. Cover the remainder of the floor, following the steps on page 60.

Draw reference lines and establish the tile layout (see page 83). Mix a batch of thin-set mortar, then spread the mortar evenly against both reference lines of one quadrant, using a ¼" square-notched trowel. Use the notched edge of the trowel to create furrows in the mortar bed.

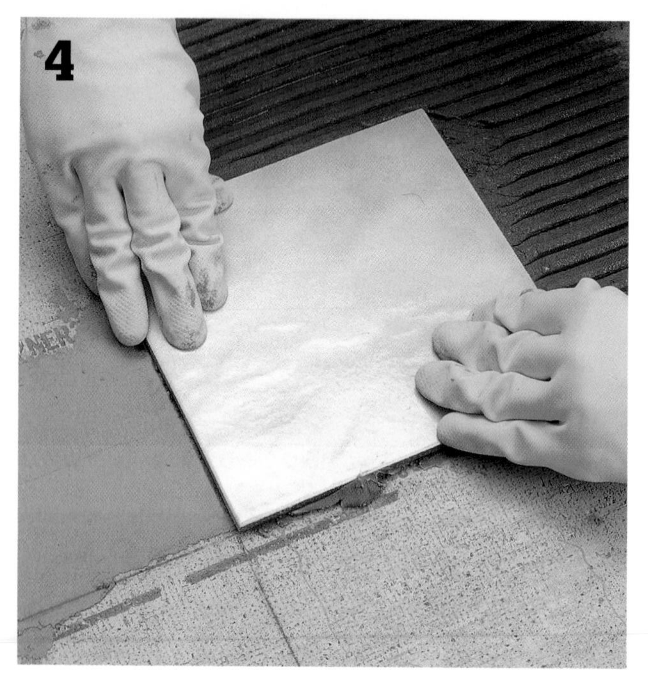

Set the first tile in the corner of the quadrant where the reference lines intersect. When setting tiles that are 8" square or larger, twist each tile slightly as you set it into position.

Using a soft rubber mallet, gently tap the central area of each tile a few times to set it evenly into the mortar.

Variation: For large tiles or uneven stone, use a larger trowel with notches that are at least ½" deep.

Variation: For mosaic sheets, use a 3/16" V-notched trowel to spread the mortar and a grout float to press the sheets into the mortar. Apply pressure gently to avoid creating an uneven surface.

To ensure consistent spacing between tiles, place plastic tile spacers at the corners of the set tile. With mosaic sheets, use spacers equal to the gaps between tiles.

(continued)

Position and set adjacent tiles into the mortar along the reference lines. Make sure the tiles fit neatly against the spacers.

To make sure the tiles are level with one another, place a straight piece of 2 × 4 across several tiles, then tap the board with a mallet.

Lay tile in the remaining area covered with mortar. Repeat steps 2 to 7, continuing to work in small sections, until you reach walls or fixtures.

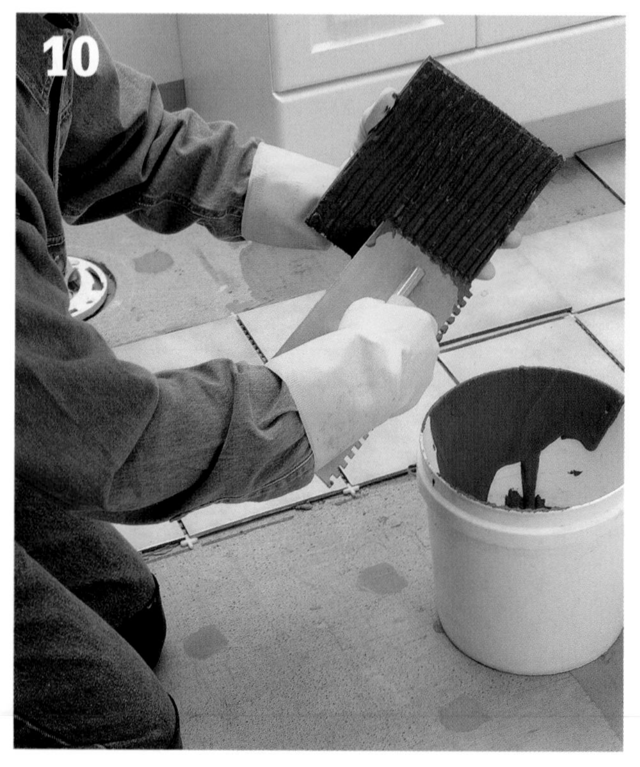

Measure and mark tiles to fit against walls and into corners (see pages 137 to 138). Cut the tiles to fit. Apply thin-set mortar directly to the back of the cut tiles, instead of the floor, using the notched edge of the trowel to furrow the mortar

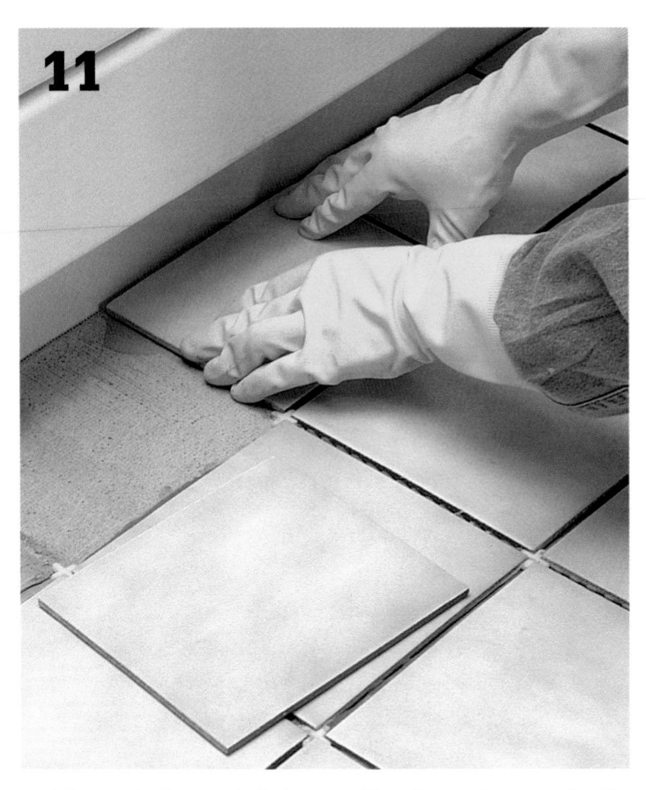

Set the cut pieces of tile into position. Press down on the tile until each piece is level with adjacent tiles.

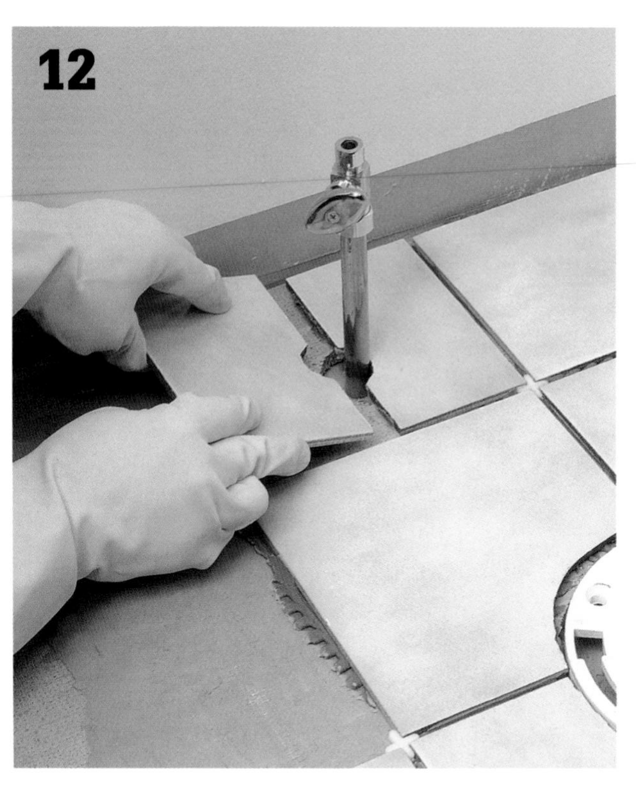

Measure, cut, and install tiles that require notches or curves to fit around obstacles, such as exposed pipes or toilet drains.

Carefully remove the spacers with needlenose pliers before the mortar hardens.

Apply mortar and set tiles in the remaining quadrants, completing one quadrant before starting the next. Inspect all of the tile joints and use a utility knife or grout knife to remove any high spots of mortar that could show through the grout.

(continued)

Install threshold material in doorways. If the threshold is too long for the doorway, cut it to fit with a jig saw or circular saw and a tungsten-carbide blade. Set the threshold in thin-set mortar so the top is even with the tile. Keep the same space between the threshold as between tiles. Let the mortar set for at least 24 hours.

Prepare a small batch of floor grout to fill the tile joints. When mixing grout for porous tile, such as quarry or natural stone, use an additive with a release agent to prevent grout from bonding to the tile surfaces.

Starting in a corner, pour the grout over the tile. Use a rubber grout float to spread the grout outward from the corner, pressing firmly on the float to completely fill the joints. For best results, tilt the float at a 60° angle to the floor and use a figure eight motion.

Use the grout float to remove excess grout from the surface of the tile. Wipe diagonally across the joints, holding the float in a near-vertical position. Continue applying grout and wiping off excess until about 25 square feet of the floor has been grouted.

Wipe a damp grout sponge diagonally over about 2 square feet of the floor at a time. Rinse the sponge in cool water between wipes. Wipe each area only once since repeated wiping can pull grout back out of joints. Repeat steps 15 to 18 to apply.

Allow the grout to dry for about 4 hours, then use a soft cloth to buff the tile surface and remove any remaining grout film.

Apply grout sealer to the grout lines, using a small sponge brush or sash brush. Avoid brushing sealer on to the tile surfaces. Wipe up any excess sealer immediately.

Variation: Use a tile sealer to seal porous tile, such as quarry tile or unglazed tile. Following the manufacturer's instructions, roll a thin coat of sealer over the tile and grout joints, using a paint roller and extension handle.

How to Install Base Trim

Dry-fit the tiles to determine the best spacing. Grout lines in base tile do not always align with grout lines in the floor tile. Use rounded bullnose tiles at outside corners, and mark tiles for cutting as needed.

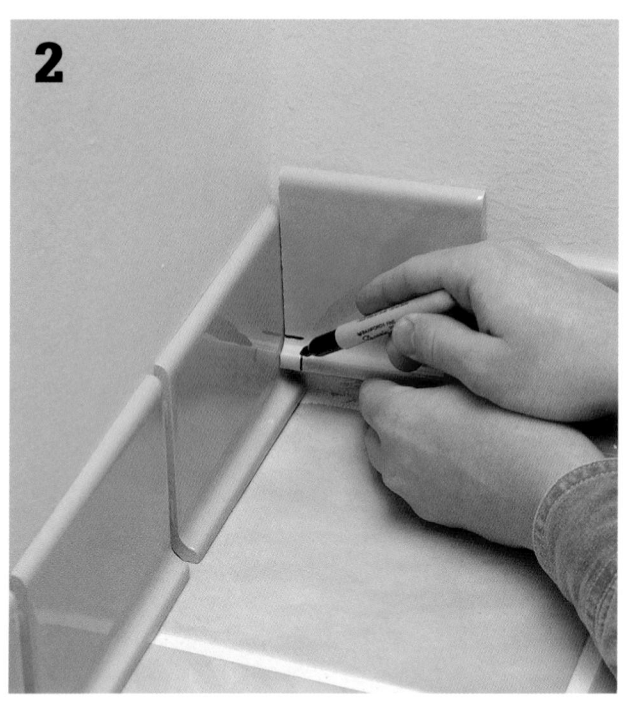

Leaving a ⅛" expansion gap between tiles at corners, mark any contour cuts necessary to allow the coved edges to fit together. Use a jig saw with a tungsten-carbide blade to make curved cuts.

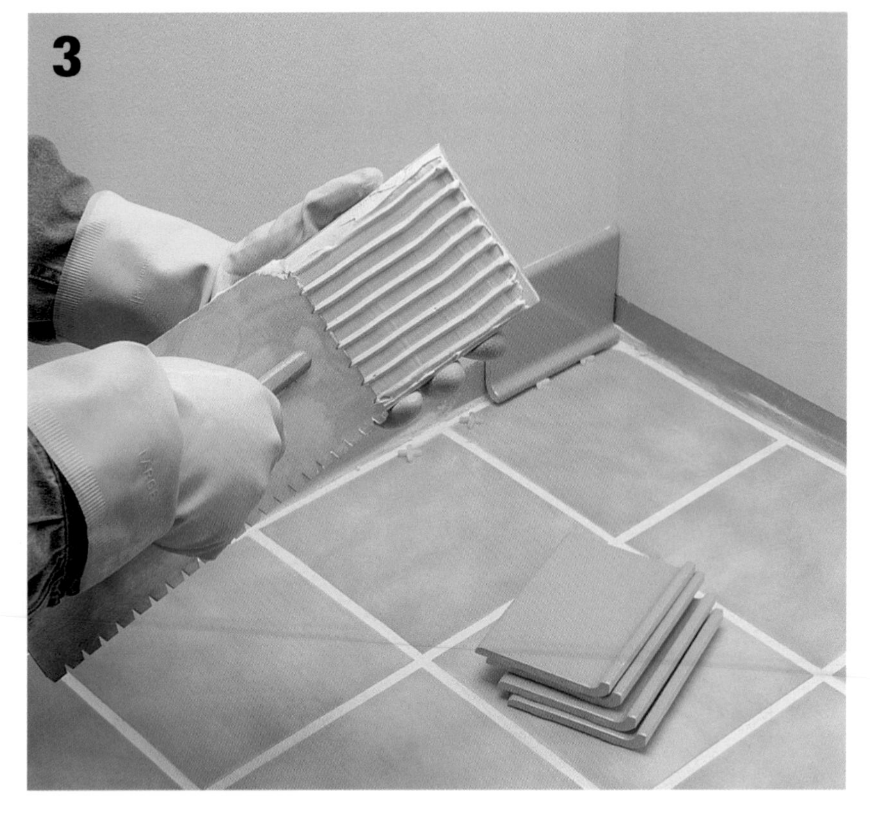

Begin installing base-trim tiles at an inside corner. Use a notched trowel to apply wall adhesive to the back of the tile. Place ⅛" spacers on the floor under each tile to create an expansion joint.

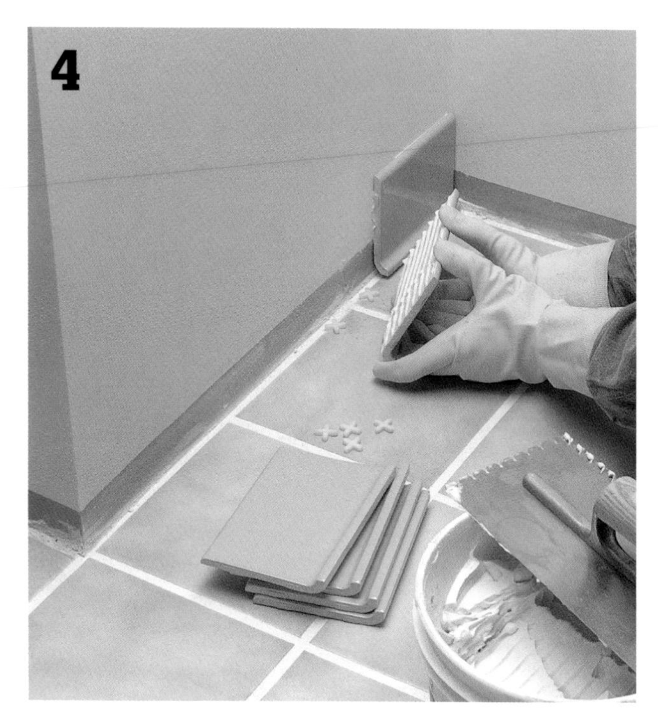

Press the tile onto the wall. Continue setting tiles, using spacers to maintain ⅛" gaps between the tiles and ⅛" expansion joints between the tiles and floor.

Use a double-bullnose tile on one side of outside corners to cover the edge of the adjoining tile.

After the adhesive dries, grout the vertical joints between tiles and apply grout along the tops of the tiles to make a continuous grout line. Once the grout hardens, fill the expansion joint between the tiles and floor with caulk.

Mosaic Glass Tile

Mosaic tile is an excellent choice for smaller areas. It requires the same preparation and handling as larger tiles, with a few differences. Sheets of mosaic tile are held together by a fabric mesh underneath. This makes them more difficult to hold, place, and move. They may not be square with your guidelines when you first lay them down. And mosaic tiles will require many more temporary spacers and much more grout.

A few cautions: Variations in color and texture are just as likely with mosaic tile as with individual tiles, so buy all your tile from the same lot and batch. Mortar or mastic intended for ceramic tile may not work with glass mosaic tile. Finally, if the finished project will be exposed to the elements, make sure you have adhesive and grout suitable for outdoor use.

Tools & Materials ▸

- Carpenter's square
- Chalk line
- Cleaning supplies
- Coarse sponge
- Craft/utility knife
- Grout sealer
- Marking pen or pencil
- Measuring tape
- Notched trowel
- Recommended adhesive
- Rubber mallet
- Sanded grout
- Scrap lumber
- Straightedge
- Tile nippers
- Tile spacers

How to Install Mosaic Tile

Clean and prepare the area as you would for individual tiles, with reference lines beginning in the center. Beginning at the center intersection, apply the recommended adhesive to one quadrant. Spread it outward evenly with a notched trowel. Lay down only as much adhesive as you can cover in 10 to 15 minutes.

Select a sheet of mosaic tile. Place several plastic spacers within the grid so that the sheet remains square. Pick up the sheet of tiles by diagonally opposite corners. This will help you hold the edges up so that you don't trap empty space in the middle of the sheet.

Gently press one corner into place on the adhesive. Slowly lower the opposite corner, making sure the sides remain square with your reference lines. Massage the sheet into the adhesive, being careful not to press too hard or twist the sheet out of position. Insert a few spacers in the outside edges of the sheet you have just placed. This will help keep the grout lines consistent.

When you have placed two or three sheets, lay a scrap piece of flat lumber across the tops and tap the wood with a rubber mallet to set the fabric mesh in the adhesive, and to force out any trapped air.

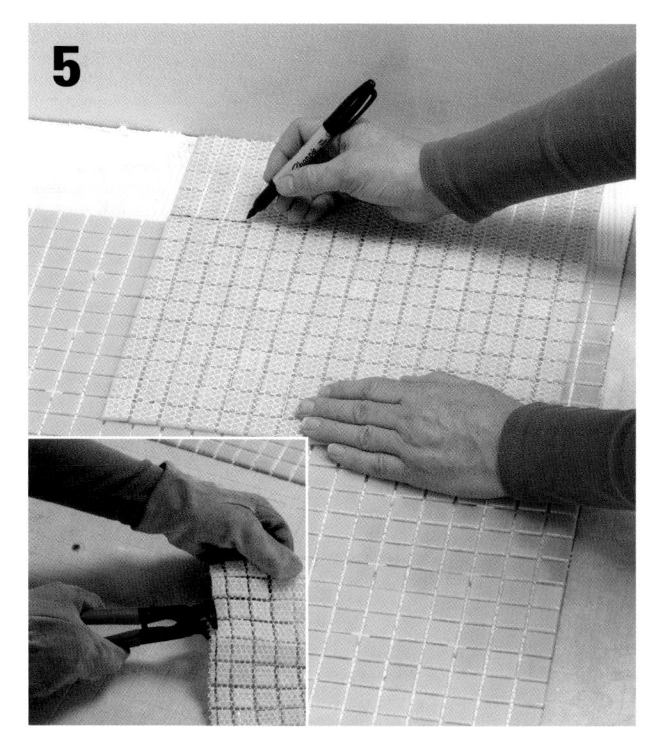

At the outer edges of your work area, you will probably need to trim one or more rows from the last sheet. If the space left at the edge is more than the width of a regular grout line, use tile nippers to trim the last row that will fit. Save these leftover tiles for repairs.

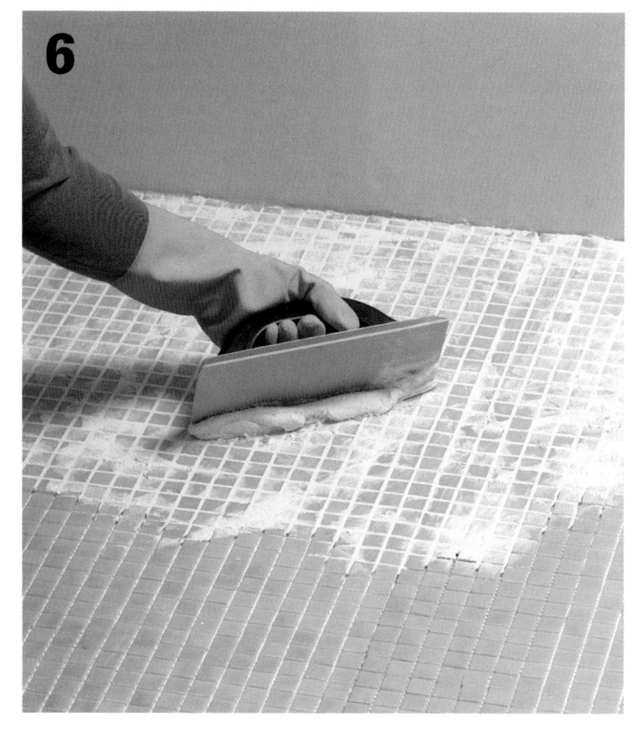

After the adhesive has cured, usually 24 to 48 hours, apply grout as you would for individual tiles. With many more spaces, mosaic tiles will require more grout. Follow the manufacturer's instructions for spreading and floating the grout. Clean up using the instructions for individual tiles (see page 155).

Installing Borders

Borders add instant appeal to any floor. They can divide a floor into sections, or they can define a particular area of flooring, such as a mosaic. You can create a design inside the border by merely turning the tile at a 45° angle or installing decorative tiles. Designs with borders should cover between a quarter and half of the floor. If the design is too small, it'll get lost in the floor. If it's too big, it'll be distracting.

You'll need to determine the size and location of your border on paper, then transfer your measurements onto the floor. A dry run with the border and field tile is still required to ensure a smooth layout.

The tile is installed in three stages. The border is placed first, followed by outside field tile, then field tile within the border.

A border catches the eye and brings a creative element to the floor. Adding a border and using different colors for the tiles within the border bring the above floor to life.

Tools & Materials ▸

- Carpenter's square
- Chalk line
- Cleaning supplies
- Coarse sponge
- Craft/utility knife
- Grout sealer
- Marking pen or pencil
- Measuring tape
- Notched trowel
- Recommended adhesive
- Rubber mallet
- Sanded grout
- Scrap lumber
- Straightedge
- Tile nippers
- Tile spacers

How to Lay Out Borders

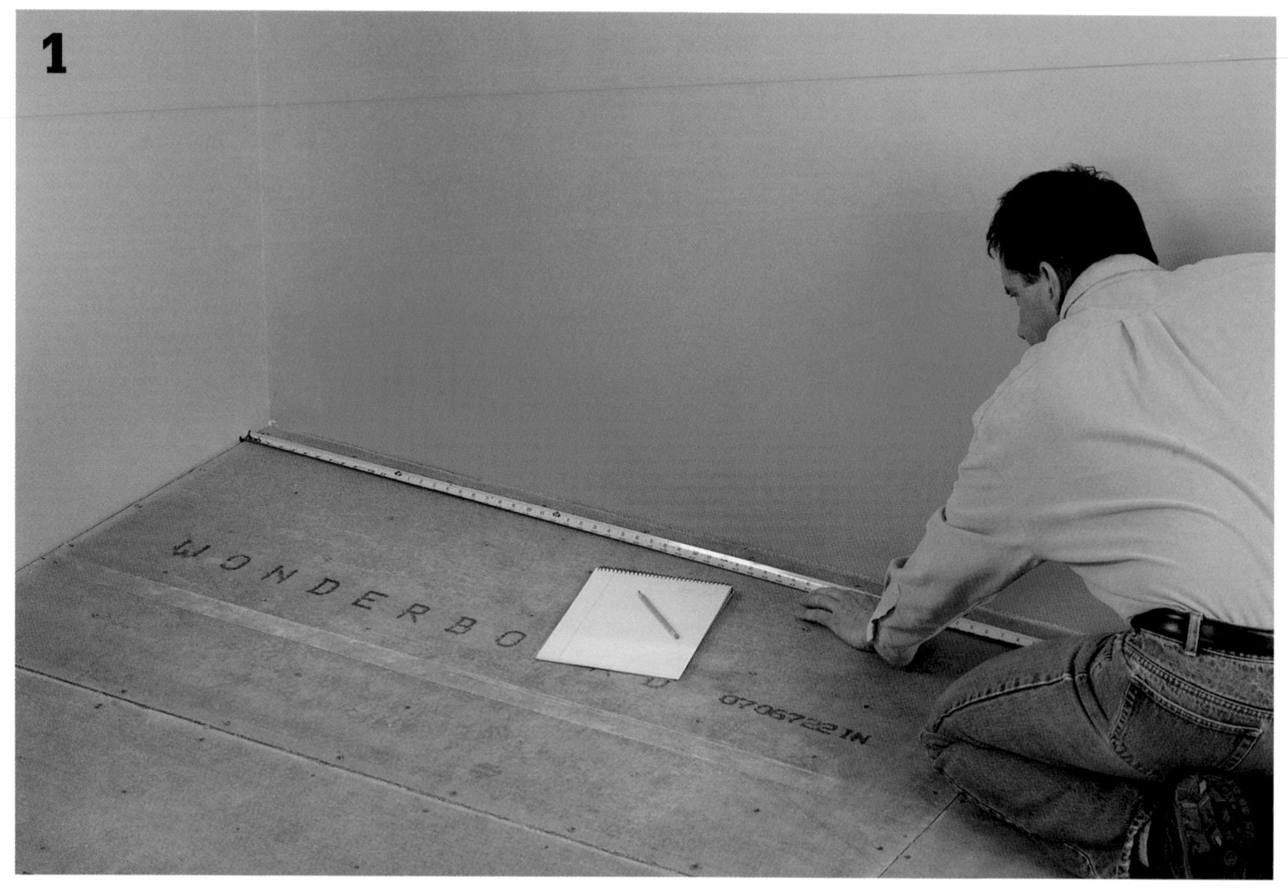

Measure the length and width of the room in which you'll be installing the border.

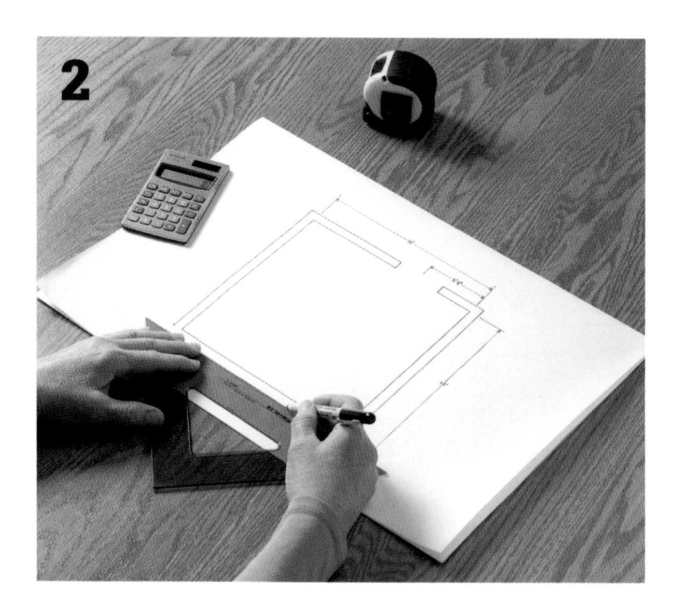

Transfer the measurements onto paper by making a scale drawing of the room. Include the locations of cabinets, doors, and furniture that will be in the room.

Determine the size of the border you want. Bordered designs should be between ¼ and ½ the area of the room. Draw the border on transparency paper, using the same scale as the room drawing.

(continued)

Place the transparency of the border over the room drawing. Move it around to find the best layout. Tape the border transparency in place over the room drawing. Draw perpendicular lines through the center of the border and calculate the distance from the center lines to the border.

Transfer the measurements from the border transparency onto your floor, starting with your center lines. Snap chalk lines to establish your layout for the border.

Lay out the border along the reference lines in a dry run. Do a dry run of the field tiles along the center lines inside and outside of the border. Make any adjustments, if necessary.

How to Lay a Running-bond Tile Pattern

Start running-bond tile by dry-fitting tile to establish working reference lines. Dry-fit a few tiles side by side using spacers. Measure the total width of the dry-fitted section. Use this measurement to snap a series of equally spaced parallel lines to help keep your tiles straight during installation. Running-bond layouts are most effective with rectangular tiles.

Starting at a point where the layout lines intersect, spread thin-set mortar and lay the first row of tiles. Offset the next row by a measurement that's equal to one-half the length of the tile plus one-half the width of the grout line.

Continue setting tiles, filling one quadrant at a time. Use the parallel reference lines as guides to keep the rows straight. Immediately wipe away any mortar that falls on the tiles. When finished, allow the mortar to cure, then grout and clean the tile (see pages 154 to 155).

How to Lay Hexagonal Tile

Snap perpendicular reference lines on the underlayment. Lay out three or four tiles in each direction along the layout lines. Place plastic spacers between the tiles to maintain even spacing. Measure the length of this layout in both directions (A and B). Use measurement A to snap a series of equally spaced parallel lines across the entire floor, then do the same for measurement B in the other direction.

Apply dry-set mortar and begin setting tile the same way as with square tile (pages 150 to 153). Apply mortar directly to the underside of any tiles that extend outside the mortar bed.

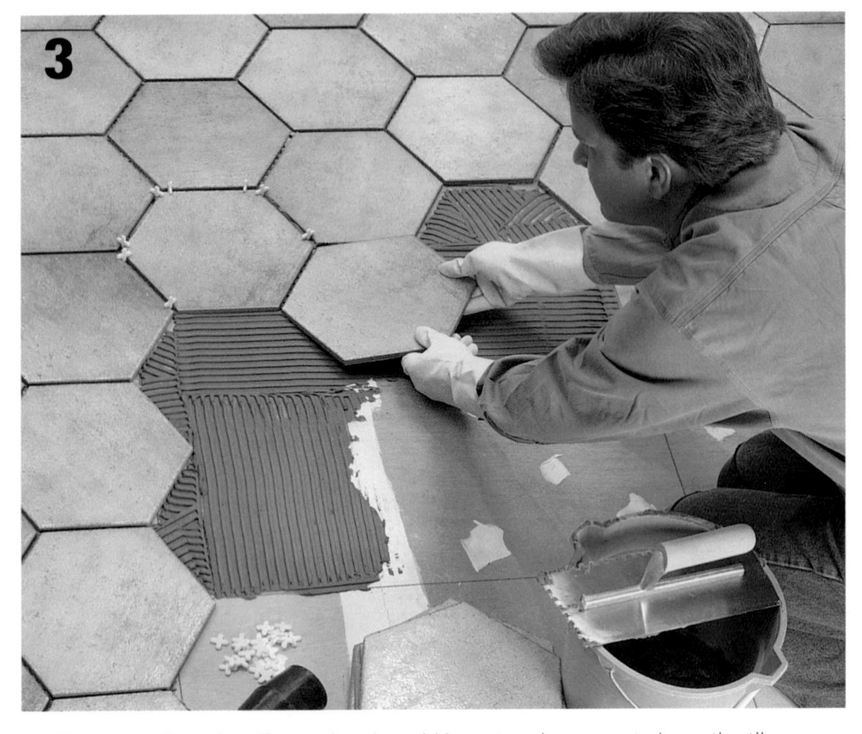

Continue setting the tiles, using the grid layout and spacers to keep the tiles aligned. Wipe off any mortar that falls onto the tile surface. When finished, allow the mortar to set, and then apply grout between tiles.

How to Lay a Diagonal Pattern with a Perpendicular Border

Follow steps 1 to 6 on pages 161 to 162 to plan your border lay out in the room. Dry-fit border tiles with spacers in the planned area. Make sure the border tiles are aligned with the reference lines. Dry-fit tiles at the outside corners of the border arrangement. Adjust the tile positions as necessary to create a layout with minimal cutting. When the layout of the tiles is set, snap chalk lines around the border tiles and trace along the edges of the outside tiles. Install the border tiles.

Draw diagonal layout lines at a 45° angle to the perpendicular reference lines.

Use standard tile-setting techniques to set field tiles inside the border. Kneel on a wide board to distribute your weight if you need to work in a tiled area that has not cured overnight.

Carpet

Carpet remains one of the most popular and versatile of all floor coverings. Almost every home has wall-to-wall carpet in at least a few rooms. It's available in an almost endless variety of colors, styles, and patterns. It can also be custom made to express a more personal design. Most carpet is nylon based, although acrylic and polyester are also popular. Wool carpeting is more formal and more expensive, but also quite popular.

Part of carpet's appeal is its soft texture. It's pleasant to walk on—especially with bare feet, since it's soft and warm underfoot—and is comfortable for children to play on. Because carpet has a pad underneath that acts as a cushion, carpet can help reduce "floor fatigue."

Carpet absorbs more noise than most other floors, thereby reducing sound between rooms. It also serves as a natural insulator and decreases heat loss through the floor. Wall-to-wall carpet can increase the R-value, or insulation level, of a room.

Carpet offers several universal design advantages. With its non-skid surface, carpet helps reduce falls, which is important for people with limited mobility. Unlike some hard floors, carpet produces no glare, which helps people with vision limitations.

Dense plush carpet creates a feeling of relaxation that's right at home in private quarters.

Shag is still fashionable, as in this living area. The carpet color matches the couch and walls to create a monochromatic palette in the room, allowing for texture to take center stage.

Buying & Estimating Carpet

When choosing carpet, one thing to consider is color and pattern. Lighter shades and colors show dirt and stains more readily, but they provide an open, spacious feel. Darker colors and multi-colored patterns don't show as much dirt or wear, but they can also make a room appear smaller.

The materials used in a carpet and its construction can affect the carpet's durability. In high traffic areas, such as hallways and entryways, a top-quality fiber will last longer. Carpet construction, the way in which fibers are attached to the backing, impacts resistance to wear and appearance.

Available widths of certain carpets may affect your buying decision; a roll that's wide enough to cover an entire room eliminates the need for seaming. When seaming is unavoidable, calculate the total square footage to be covered, then add 20 percent to cover trimming and seaming.

The type of carpet will dictate the type of pad you should use. Check carpet sample labels for the manufacturer's recommendations. Since carpet and padding work in tandem to create a floor covering system, use the best pad you can afford that works with your carpet. In addition to making your carpet feel more plush underfoot, the pad makes your floor quieter and warmer. A high-quality pad also helps reduce carpet wear.

Labels on the back of samples usually tell you the fiber composition, the available widths (usually 12 or 15 feet), what anti-stain treatments and other finishes were applied, and details of the product warranty.

Tips for Evaluating Carpet ▸

Fiber Type	Characteristics
Nylon	Easy to clean, very durable, good stain resistance; colors sometimes fade in direct sunlight.
Polyester	Excellent stain resistance, very soft in thick cut-pile constructions; colors don't fade in sunlight.
Olefin	Virtually stain- and fade-proof, resists moisture and static; not as resilient as nylon or as soft as polyester.
Acrylic	Resembles wool in softness and look, good moisture resistance; less durable than other synthetics.
Wool	Luxurious look and feel, good durability and warmth; more costly and less stain-resistant than synthetics.

Carpet Construction

The top surface of a carpet, called the pile, consists of yarn loops pushed up through a backing material. The loops are left intact or cut by the manufacturer, depending on the desired effect. Most carpet sold today is made from synthetic fibers, such as nylon, polyester, and olefin, although natural wool carpet is still popular.

A good rule of thumb for judging the quality of a carpet is to look at the pile density. Carpet with many pile fibers packed into a given area will resist crushing, repel stains and dirt buildup better, and be more durable than carpet with low pile density.

Cushion-backed carpet has a foam backing bonded to it, eliminating the need for additional padding. Cushion-backed carpet is easy to install because it does not require stretching or tackless strips. Instead, it is secured to the floor with general-purpose adhesive, much like full-spread sheet vinyl. Cushion-backed carpet usually costs less than conventional carpet, but it's generally a lower-quality product.

Loop-pile carpet has a textured look created by the rounded ends of the uncut yarn loops pushed up through the backing. The loops can be arranged randomly or they can make a distinct pattern, such as herringbone. Loop pile is ideal for heavy-traffic areas since loops are virtually impervious to crushing.

Velvet cut-pile carpet has the densest pile of any carpet type. It's cut so the color remains uniform when the pile is brushed in any direction. Velvets are well suited to formal living spaces.

Saxony cut-pile carpet, also known as plush, is constructed to withstand crushing and matting better than velvets. The pile is trimmed at a bevel, giving it a speckled appearance.

Labels on the back of samples usually tell you the fiber composition, the available widths (usually 12 or 15 feet), what anti-stain treatments and other finishes were applied, and details of the product warranty.

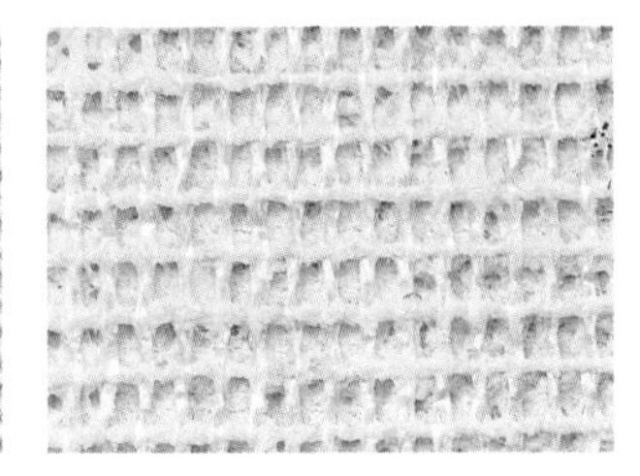

Examine the backing or "foundation" of the carpet. A tighter grid pattern (left) usually indicates dense-pile carpet that will be more durable and soil-resistant than carpet with looser pile (right).

Planning & Layout ▸

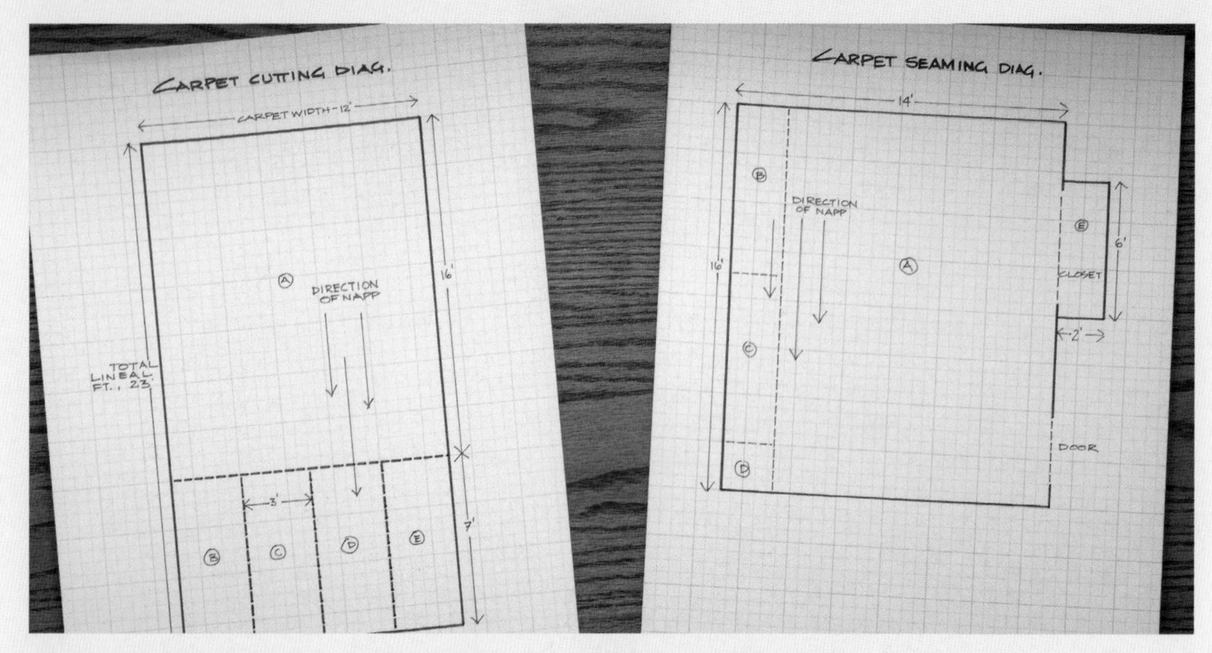

Sketch a scale drawing of the factory carpet roll and another drawing of the room to be carpeted. Use the drawings to plan the cuts and determine how the carpet pieces will be arranged. In most large rooms, the installation will include one large piece of carpet the same width as the factory roll and several smaller pieces that are permanently seamed to the larger piece. When sketching the layout, remember that carpet pieces must be oversized to allow for precise seaming and trimming. Your finished drawings will tell you the length of carpet you need to buy.

Keep pile direction consistent. Carpet pile is usually slanted, which affects how the carpet looks from different angles as light reflects off the surface. Place seamed pieces so the pile faces the same direction.

Maintain patterns when seaming patterned carpet. Because of this necessity, there's always more waste when installing patterned carpet. For a pattern that repeats itself every 18", for example, each piece must be oversized 18" to ensure the pattern is aligned. Pattern repeat measurements are noted on carpet samples.

At seams, add an extra 3" to each piece when estimating the amount of carpet you'll need. This extra material helps when cutting straight edges for seaming.

At each wall, add 6" for each edge that's along the wall. This surplus will be trimmed away when the carpet is cut to the exact size of the room.

Closet floors are usually covered with a separate piece of carpet that's seamed to the carpet in the main room area.

For stairs, add together the rise and run of each step to estimate the carpet needed for the stairway. Measure the width of the stairway to determine how many strips you can cut from the factory roll. For a 3 ft.-wide stairway, for example, you can cut three strips from a 12 ft.-wide roll, allowing for waste. Rather than seaming carpet strips together end to end, plan the installation so the ends of the strips fall in the stair crotches (see page 197). When possible, try to carpet stairs with a single carpet strip.

Carpet Tools & Materials

Installing carpet requires the use of some specialty tools, most notably the knee kicker and power stretcher. These tools are available at most rental centers and carpet stores.

Other than the carpet itself, the pad is the most important material in carpet installation. In addition to making your carpet feel more comfortable, it helps reduce sounds. The pad also helps keep warm air from escaping through your floor, thereby keeping the carpet warmer.

By cushioning the carpet fibers, the pad reduces wear and extends the life of your carpet. Be sure to use a quality pad.

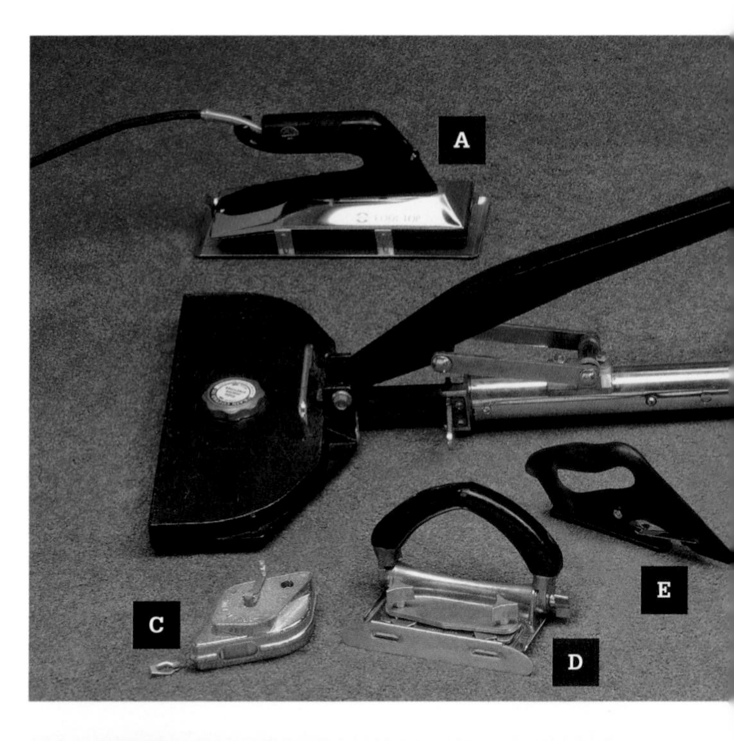

Carpeting tools include: seam iron (A), power stretcher and extensions (B), chalk line (C), edge trimmer (D), row-running knife (E), utility knife (F), stair tool (G), hammer (H), knee kicker (I), aviation snips (J), scissors (K), and stapler (L).

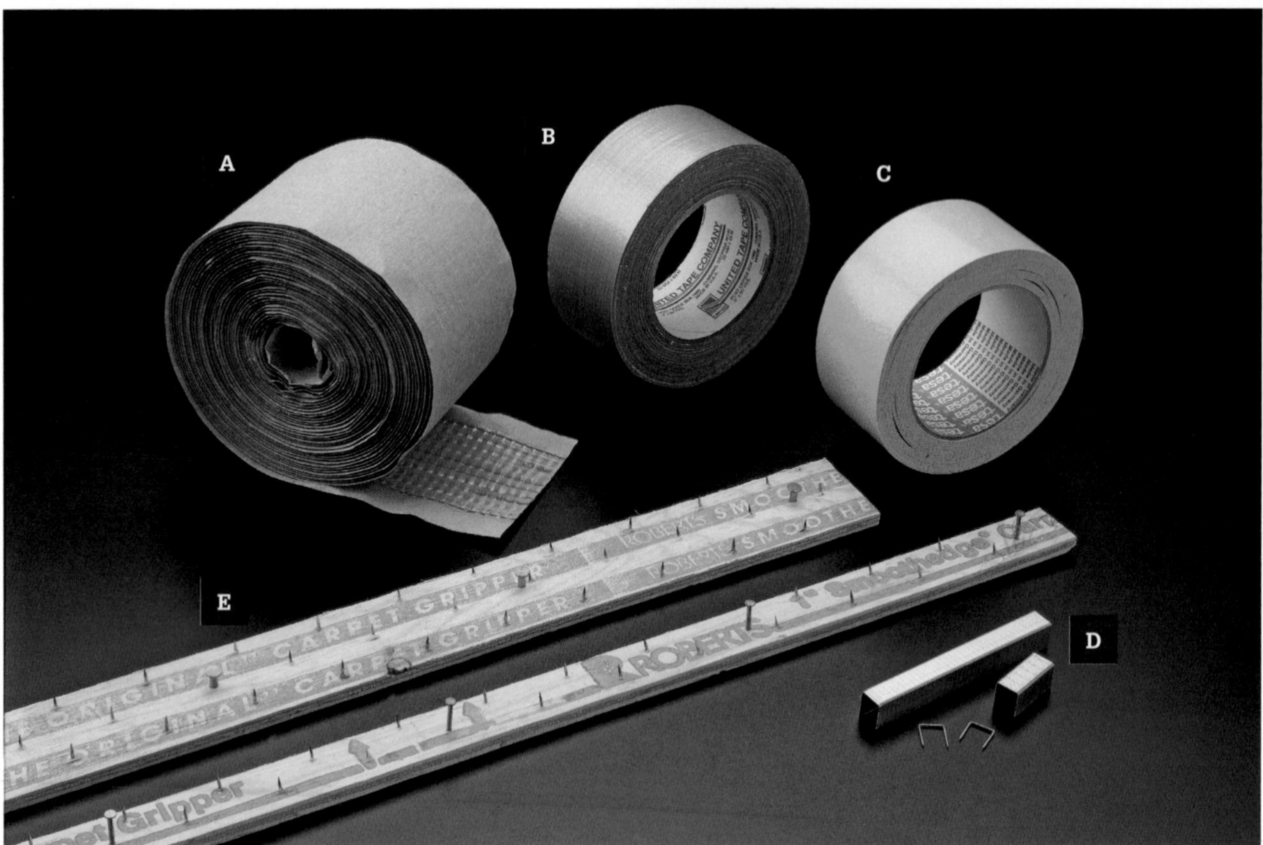

Carpeting materials include: hot-glue seam tape (A), used to join carpet pieces together; duct tape (B), for seaming carpet pads; double-sided tape (C), used to secure carpet pads to concrete; staples (D), used to fasten padding to underlayment; and tackless strips (E), for securing the edges of stretched carpet.

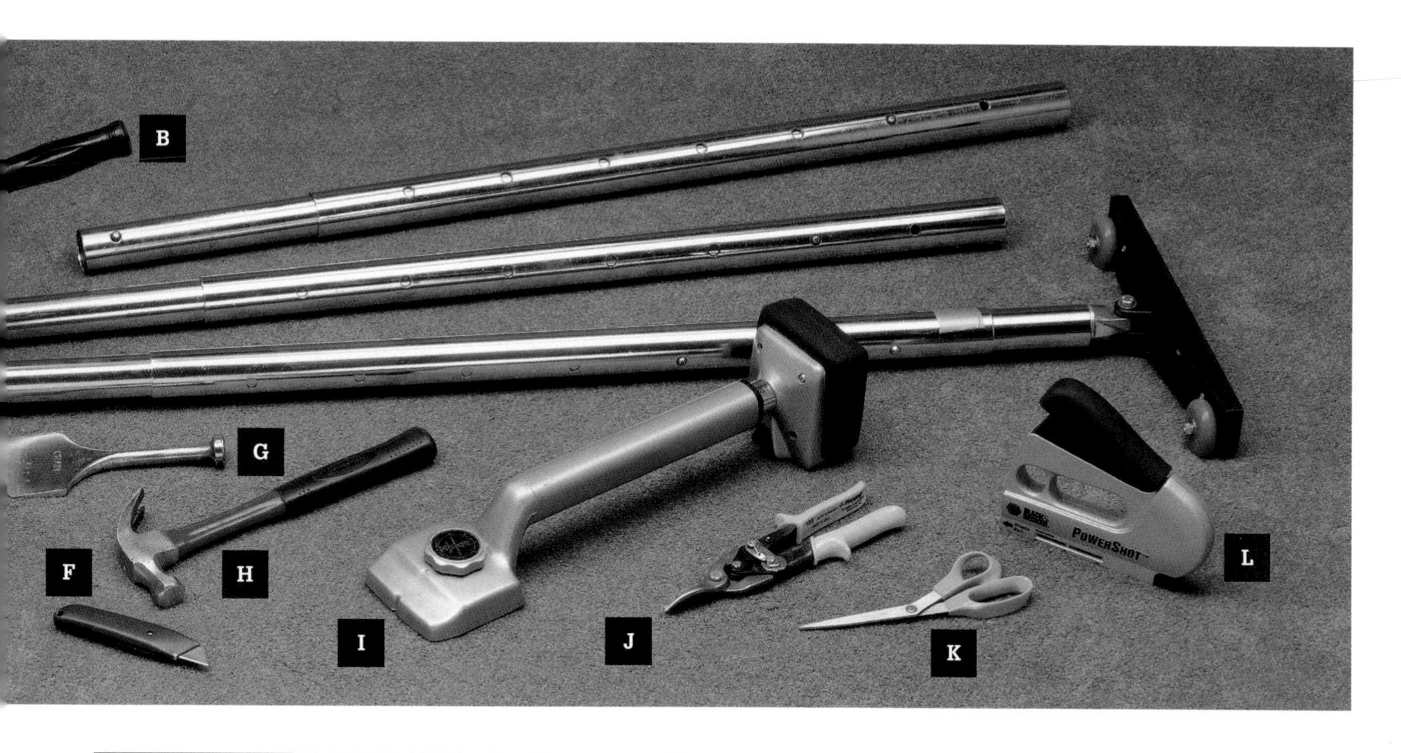

Carpet padding comes in several varieties, including: bonded urethane foam (A), cellular sponge rubber (B), grafted prime foam (C), and prime urethane (D). Bonded urethane padding is suitable for low-traffic areas, while prime urethane and grafted prime foam are better for high-traffic areas. In general, cut pile, cut-and-loop, and high-level loop carpets perform best with prime or bonded urethane or rubber pads that are less than 7⁄16" thick. For Berbers or other stiff-backed carpets, use 3⁄8"-thick bonded urethane foam or cellular sponge rubber. Foam padding is graded by density: the denser the foam, the better the pad. Rubber padding is graded by weight: the heavier, the better.

Using Carpet Tools

The knee kicker and power stretcher are the two most important tools for installing carpet. They are used to stretch a carpet smooth and taut before securing it to tackless strips installed around the perimeter of a room.

The power stretcher is the more efficient of the two tools and should be used to stretch and secure as much of the carpet as possible. The knee kicker is used to secure carpet in tight areas where the power stretcher can't reach, such as closets.

A logical stretching sequence is essential to good carpet installation. Begin by attaching the carpet at a doorway or corner, then use the power stretcher and knee kicker to stretch the carpet away from the attached areas and toward the opposite walls.

How to Use a Knee Kicker

A knee kicker (and power stretcher) has teeth that grab the carpet foundation for stretching. Adjust the depth of the teeth by turning the knob on the knee kicker head. The teeth should be set deep enough to grab the carpet foundation without penetrating to the padding.

Place the kicker head a few inches away from the wall to avoid dislodging the tackless strips, then strike the kicker cushion firmly with your knee, stretching the carpet taut. Tack the carpet to the pins on the tackless strips to hold it in place.

How to Use a Power Stretcher

Align the pieces of the power stretcher along the floor with the tail positioned at a point where the carpet is already secured and the head positioned just short of the opposite wall. Fit the ends of the sections together.

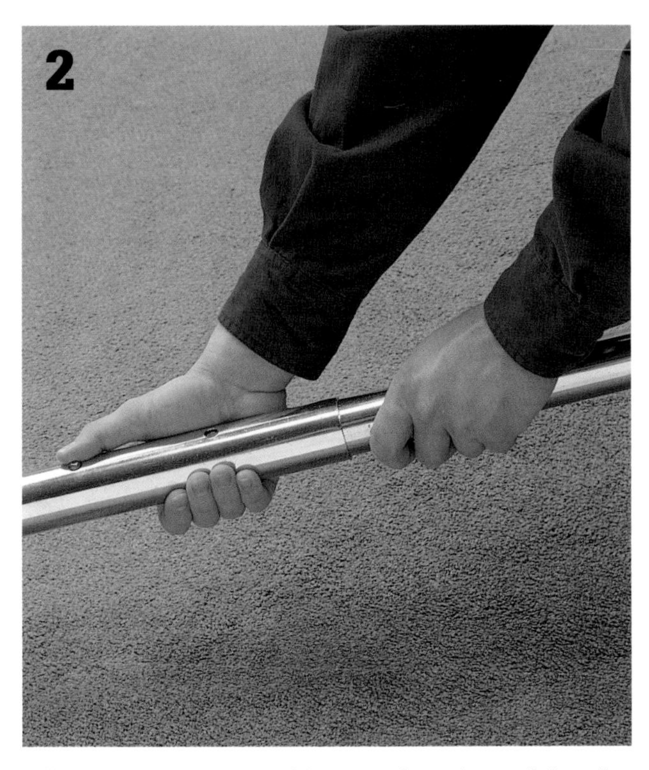

Telescope one or more of the extension poles until the tail rests against the starting wall or block and the head is about 5" from the opposite wall.

Adjust the teeth on the head so they grip the carpet foundation (see step 1, opposite page). Depress the lever on the head to stretch the carpet. The stretcher head should move the carpet about 2".

Installing Carpet Transitions

Doorways, entryways, and other transition areas require special treatment when installing carpet. Transition materials and techniques vary, depending on the level and type of the adjoining flooring (see photos, right).

For a transition to a floor that's either at the same height or lower than the bottom of the carpet, attach a metal carpet bar to the floor and secure the carpet inside the bar. This transition is often used where carpet meets a vinyl or tile floor. Carpet bars are sold in standard door-width lengths and in longer strips.

For a transition to a floor that's higher than the carpet bottom, use tackless strips, as if the adjoining floor surface was a wall. This transition is common where carpet meets a hardwood floor.

For a transition to another carpet of the same height, join the two carpet sections together with hot-glue seam tape.

For a transition in a doorway between carpets of different heights or textures, install tackless strips and a hardwood threshold. Thresholds are available predrilled and ready to install with screws.

Tools & Materials ▸

- Hacksaw
- Marker
- Utility knife
- Knee kicker
- Stair tool
- Straightedge
- Screwdriver
- Transition materials
- Wood block

Metal carpet bar

Tackless strip tuck-under

Hot-glue seam tape

Hardwood threshold

How to Make Transitions with Metal Carpet Bars

Measure and cut a carpet bar to fit the threshold, using a hacksaw. Nail the carpet bar in place. In doorways, the upturned metal flange should lie directly under the center of the door when it's closed.

Roll out, cut, and seam the carpet. Fold the carpet back in the transition area, then mark it for trimming. The edge of the carpet should fall ⅛" to ¼" short of the corner of the carpet bar so it can be stretched into the bar.

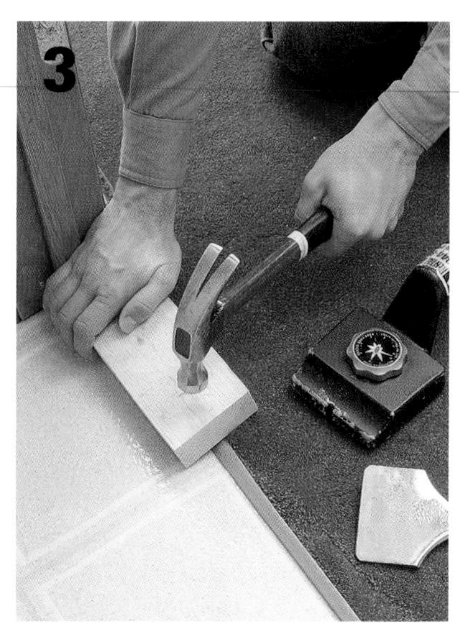

Use a knee kicker to stretch the carpet snugly into the corner of the carpet bar. Press the carpet down onto the pins with a stair tool. Bend the carpet bar flange down over the carpet by striking it with a hammer and a block of wood.

How to Make Transitions with Tackless Strips

Install a tackless strip, leaving a gap equal to ⅔ the thickness of the carpet for trimming. Roll out, cut, and seam the carpet. Mark the edge of the carpet between the tackless strip and the adjoining floor surface about ⅛" past the point where it meets the adjacent floor.

Use a straightedge and utility knife to trim the excess carpet. Stretch the carpet toward the strip with a knee kicker, then press it onto the pins of the tackless strip.

Tuck the edge of the carpet into the gap between the tackless strip and the existing floor, using a stair tool.

Installing Carpet

The easiest way to secure carpeting is to install tackless strips around the perimeter of the room. Once the strips are installed, carpet padding is rolled out as a foundation for the carpet.

Standard ¾"-wide tackless strips are adequate for securing most carpet. For carpets laid on concrete, use wider tackless strips that are attached to the concrete with masonry nails. Be careful when handling the tackless strips, since the sharp pins can be dangerous. Where the carpet meets a doorway or another type of flooring, install the appropriate transitions.

How to Install Tackless Strips

Starting in a corner, nail tackless strips to the floor, keeping a gap between the strips and the walls that's about ⅔ the thickness of the carpet. Use plywood spacers. Angled pins on the strip should point toward the wall.

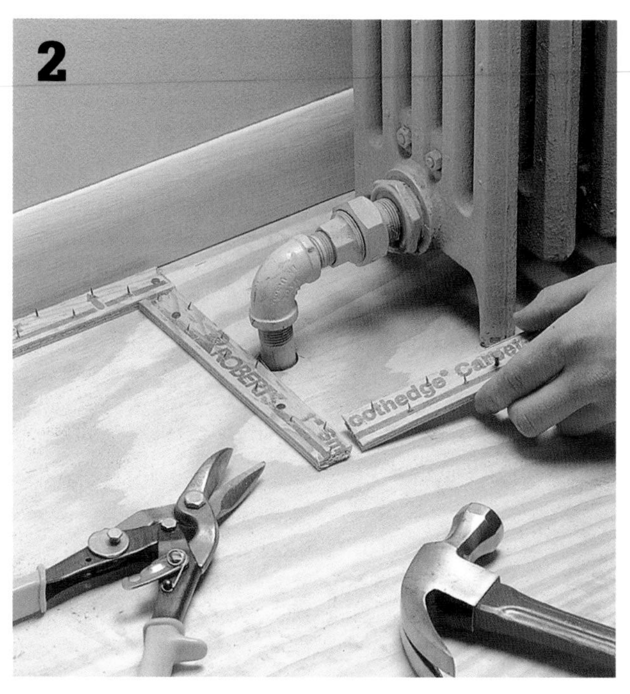

Use aviation snips to cut tackless strips to fit around radiators, door moldings, and other obstacles.

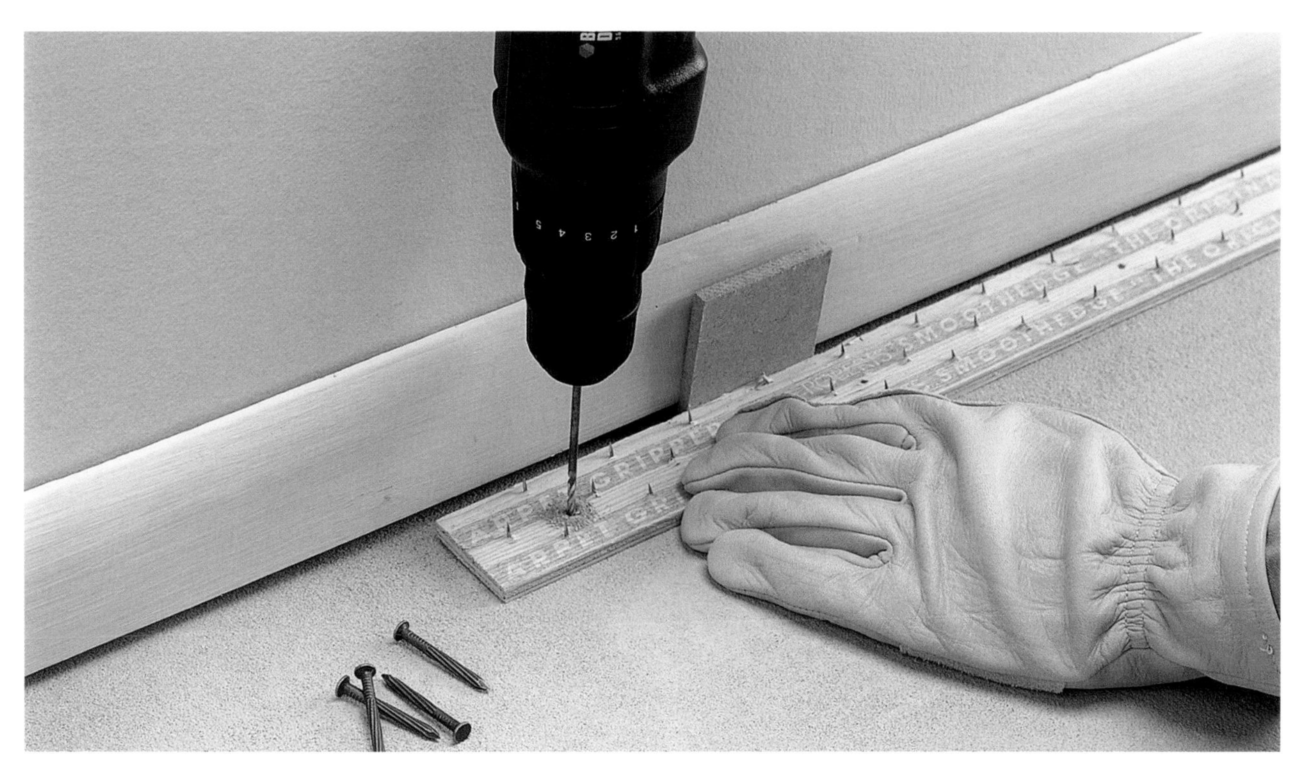

Variation: On concrete, use wider tackless strips. Using a masonry bit, drill pilot holes through the strips and into the floor. Then fasten the strips with 1½" fluted masonry nails.

How to Install Carpet Padding

Roll out enough padding to cover the entire floor. Make sure the seams between the padding are tight. If one face of the padding has a slicker surface, keep the slick surface face up, making it easier to slide the carpet over the pad during installation.

Use a utility knife to cut away excess padding along the edges. The padding should touch, but not overlap, the tackless strips.

Tape the seams together with duct tape, then staple the padding to the floor every 12".

Variation: To fasten padding to a concrete floor, apply double-sided tape next to the tackless strips, along the seams, and in an "X" pattern across the floor.

How to Cut & Seam Carpet

Position the carpet roll against one wall, with its loose end extending up the wall about 6", then roll out the carpet until it reaches the opposite wall.

At the opposite wall, mark the back of the carpet at each edge about 6" beyond the point where the carpet touches the wall. Pull the carpet back away from the wall so the marks are visible.

Variation: When cutting loop-pile carpet, avoid severing the loops by cutting it from the top side, using a row-running knife. Fold the carpet back along the cut line to part the pile (left) and make a crease along the part line. Lay the carpet flat and cut along the part in the pile (right). Cut slowly to ensure a smooth, straight cut.

(continued)

Snap a chalk line across the back of the carpet between the marks. Place a scrap piece of plywood under the cutting area to protect the carpet and padding from the knife blade. Cut along the line, using a straightedge and utility knife.

Next to walls, straddle the edge of the carpet and nudge it with your foot until it extends up the wall by about 6" and is parallel to the wall.

At the corners, relieve buckling by slitting the carpet with a utility knife, allowing the carpet to lie somewhat flat. Make sure that corner cuts do not cut into usable carpet.

Using your seaming plan as a guide, measure and cut fill-in pieces of carpet to complete the installation. Be sure to include a 6" surplus at each wall and a 3" surplus on each edge that will be seamed to another piece of carpet. Set the cut pieces in place, making sure the pile faces in the same direction on all pieces.

Roll back the large piece of carpet on the side to be seamed, then use a chalk line to snap a straight seam edge about 2" from the factory edge. Keep the ends of the line about 18" from the sides of the carpet where the overlap onto the walls causes the carpet to buckle.

Using a straightedge and utility knife, carefully cut the carpet along the chalk line. To extend the cutting lines to the edges of the carpet, pull the corners back at an angle so they lie flat, then cut the line with the straightedge and utility knife. Place scrap wood under the cutting area to protect the carpet while cutting.

On smaller carpet pieces, cut straight seam edges where the small pieces will be joined to one another. Don't cut the edges that will be seamed to the large carpet piece until after the small pieces are joined together.

(continued)

Option: Apply a continuous bead of seam glue along the cut edges of the backing at seams to ensure that the carpet will not fray.

Plug in the seam iron and set it aside to heat up, then measure and cut hot-glue seam tape for all seams. Begin by joining the small fill-in pieces to form one large piece. Center the tape under the seam with the adhesive side facing up.

Set the iron under the carpet at one end of the tape until the adhesive liquifies, usually about 30 seconds. Working in 12" sections, slowly move the iron along the tape, letting the carpet fall onto the hot adhesive behind it. Set weights at the end of the seam to hold the pieces in place.

Press the edges of the carpet together into the melted adhesive behind the iron. Separate the pile with your fingers to make sure no fibers are stuck in the glue and the seam is tight, then place a weighted board over the seam to keep it flat while the glue sets.

Variation: To close any gaps in loop-pile carpet seams, use a knee kicker to gently push the seam edges together while the adhesive is still hot.

Continue seaming the fill-in pieces together. When the tape adhesive has dried, turn the seamed piece over and cut a fresh seam edge as done in steps 7 and 8. Reheat and remove about 1½" of tape from the end of each seam to keep it from overlapping the tape on the large piece.

(continued)

Use hot-glue seam tape to join the seamed pieces to the large piece of carpet, repeating steps 10 through 12.

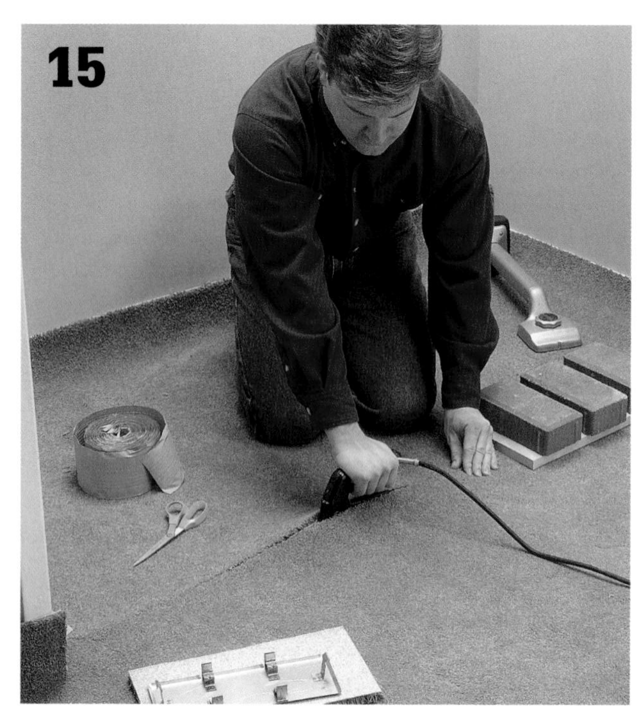

If you're laying carpet in a closet, cut a fill-in piece and join it to the main carpet with hot-glue seam tape.

Tip ▸

At radiators, pipes, and other obstructions, cut slits in the carpet with a utility knife. Cut long slits from the edge of the carpet to the obstruction, then cut short cross-slits where the carpet will fit around the obstruction.

Tip ▸

To fit carpet around partition walls where the edges of the wall or door jamb meet the floor, make diagonal cuts from the edge of the carpet at the center of the wall to the points where the edges of the wall meet the floor.

How to Stretch & Secure Carpet

Before stretching the seamed carpet, read through this entire section and create a stretching sequence similar to the one shown here. Start by fastening the carpet at a doorway threshold using carpet transitions (see pages 182 to 183).

If the doorway is close to a corner, use the knee kicker to secure the carpet to the tackless strips between the door and the corner. Also secure a few feet of carpet along the adjacent wall, working toward the corner.

Use a power stretcher to stretch the carpet toward the wall opposite the door. Brace the tail with a length of 2 × 4 placed across the doorway. Leaving the tail in place and moving only the stretcher head, continue stretching and securing the carpet along the wall, working toward the nearest corner in 12" to 24" increments.

(continued)

As you stretch the carpet, secure it onto the tackless strips with a stair tool and hammer.

With the power stretcher still extended from the doorway to the opposite side of the room, knee-kick the carpet onto the tackless strips along the closest wall, starting near the corner closest to the stretcher tail. Disengage and move the stretcher only if it's in the way.

Reposition the stretcher so its tail is against the center of the wall you just secured. Stretch and secure the carpet along the opposite wall, working from the center toward a corner. If there's a closet in an adjacent wall, work toward that wall, not the closet.

Use the knee kicker to stretch and secure the carpet inside the closet (if any). Stretch and fasten the carpet against the back wall first, then do the side walls. After the carpet in the closet is stretched and secured, use the knee kicker to secure the carpet along the walls next to the closet. Disengage the power stretcher only if it's in the way.

Return the head of the power stretcher to the center of the wall. Finish securing carpet along this wall, working toward the other corner of the room.

Reposition the stretcher to secure the carpet along the last wall of the room, working from the center toward the corners. The tail block should be braced against the opposite wall.

(continued)

Tip ▸

Locate any floor vents under the stretched carpet, then use a utility knife to cut away the carpet, starting at the center. It's important that this be done only after the stretching is complete.

10

Use a carpet edge trimmer to trim surplus carpet away from the walls. At corners, use a utility knife to finish the cuts.

11

Tuck the trimmed edges of the carpet neatly into the gaps between the tackless strips and the walls, using a stair tool and hammer.

Installing Carpet on Stairs

Where practical, try to carpet stairs with a single strip of carpet. If you must use two or more pieces, plan the layout so the pieces meet where a riser meets a tread. Do not seam carpet pieces together in the middle of a tread or riser.

The project shown here involves a staircase that's enclosed on both sides. For open staircases, turn down the edges of the carpet and secure them with carpet tacks.

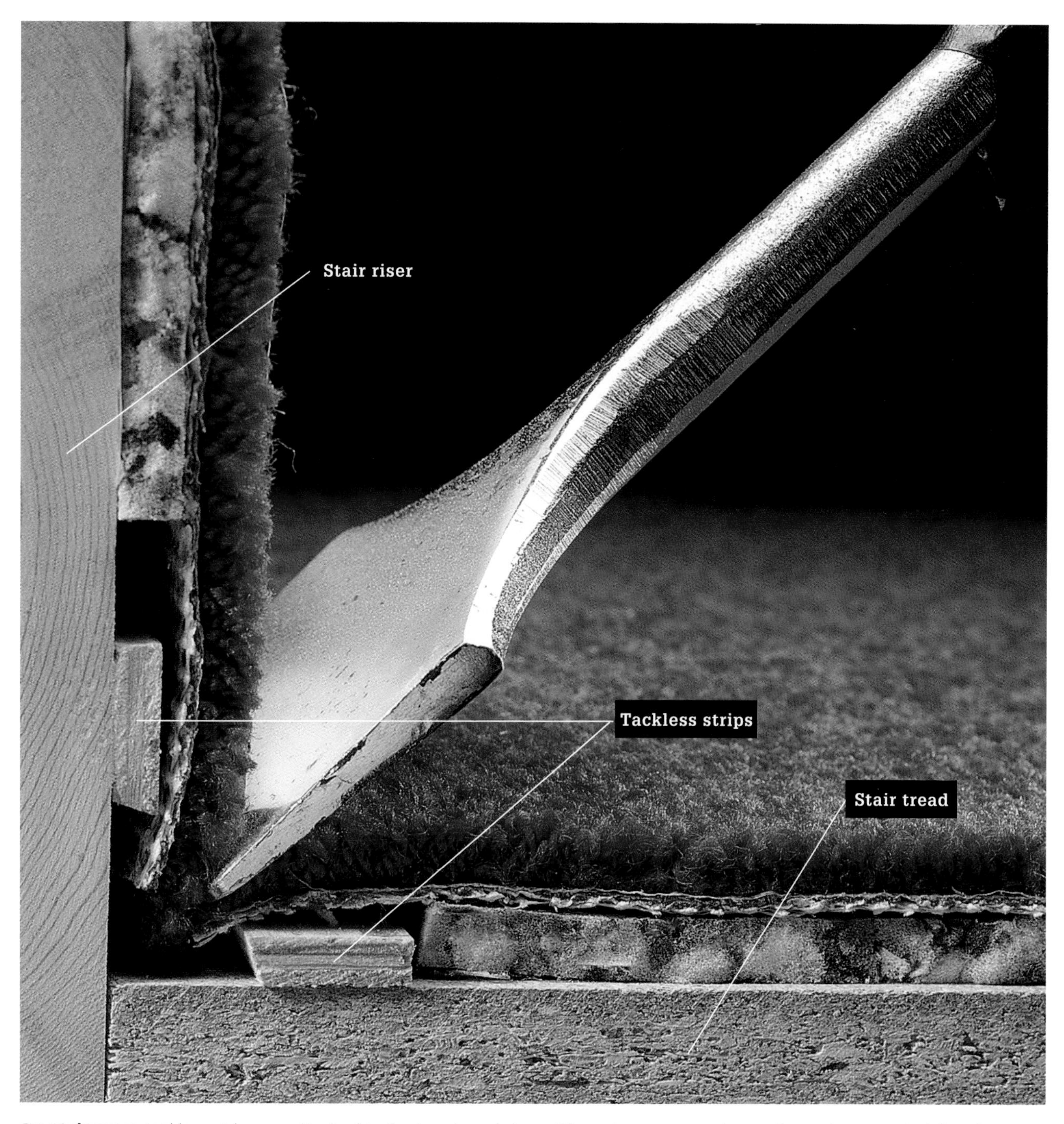

On stairways, tackless strips are attached to the treads and risers. Where two or more pieces of carpet are needed, the pieces should meet at the "crotch" of the step, where the riser and tread meet.

How to Carpet Stairs

Measure the width of the stairway. Add together the vertical rise and horizontal run of the steps to determine how much carpet you'll need. Use a straightedge and utility knife to carefully cut the carpet to the correct dimensions.

Fasten tackless strips to the risers and the treads. On the risers, place the strips about 1" above the treads. On the treads, place the strips about ¾" from the risers. Make sure the pins point toward the crotch of the step. On the bottom riser, leave a gap equal to ⅔ the carpet thickness.

For each step, cut a piece of carpet padding the width of the step and long enough to cover the tread and a few inches of the riser below it. Staple the padding in place.

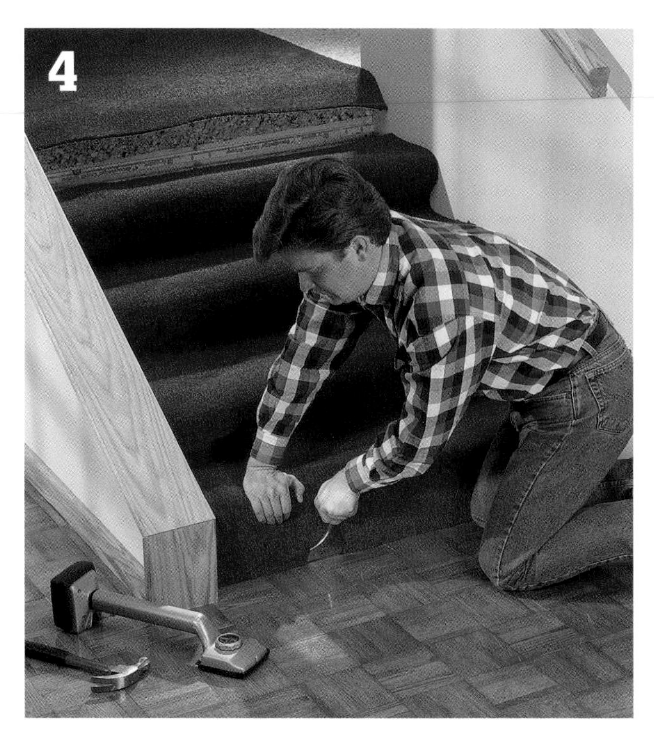

Position the carpet on the stairs with the pile direction pointing down. Secure the bottom edge using a stair tool to tuck the end of the carpet between the tackless strip and the floor.

Use a knee kicker and stair tool to stretch the carpet onto the tackless strip on the first tread. Start in the center of the step, then alternate kicks on either side until the carpet is completely secured on the step.

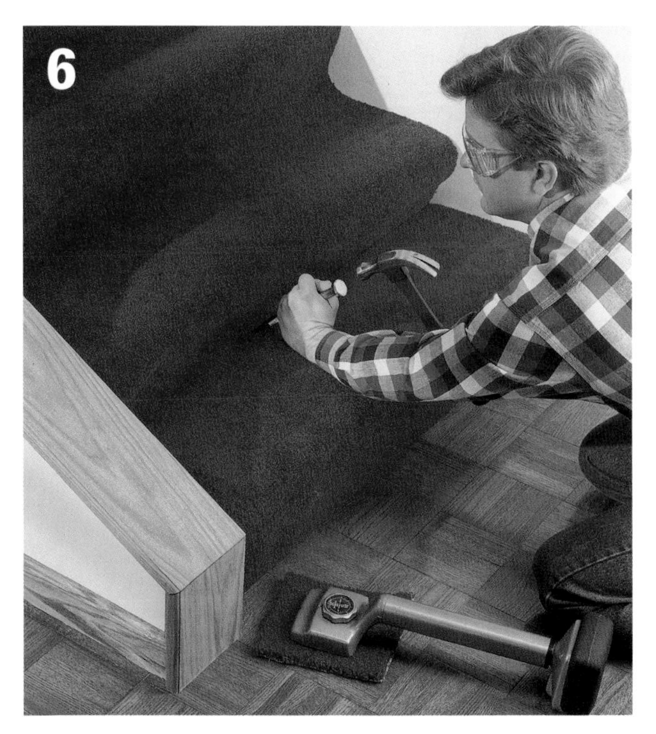

Use a hammer and stair tool to wedge the carpet firmly into the back corner of the step. Repeat this process for each step.

Where two carpet pieces meet, secure the edge of the upper piece first, then stretch and secure the lower piece.

Carpet Squares

Most carpeting has a single design and is stretched from wall to wall. It covers more square feet of American homes than any other material. You can install it yourself, following the instructions on pages 178 to 190. But if you want a soft floor covering that gives you more options, carpet squares are an excellent choice.

Manufacturers have found ways to create attractive new carpet using recycled fibers. This not only reuses material that would otherwise become landfill, it reduces waste in manufacturing as well. So, instead of adding to problems of resource consumption and pollution, carpet squares made from recycled materials help reduce them.

The squares are attached to each other and to the floor with adhesive dots. They can be installed on most clean, level, dry underlayment or existing floor. If the surface underneath is waxed or varnished, check with the manufacturer before you use any adhesives on it.

Tools & Materials ▸

- Adhesive
- Aviator's snips
- Carpenter's square
- Chalk line
- Cleaning supplies
- Craft/utility knife
- Flat-edged trowel
- Marking pen or pencil
- Measuring tape
- Notched trowel
- Straightedge

How to Install Carpet Squares

Take the squares out of the package. Usually, you want to keep new flooring out of the way until you're ready to install it. But some materials, such as carpet or sheet vinyl, should be at room temperature for at least 12 hours before you lay them down.

Check the requirements for the recommended adhesive. You can install carpet squares over many other flooring materials, including hardwood, laminates, and resilient sheets or tiles. The carpet squares shown here are fastened with adhesive dots, so almost any existing floor will provide a usable surface.

Make sure the existing floor is clean, smooth, stable, and dry. Use floor leveler if necessary to eliminate any hills or valleys. If any part of the floor is loose, secure it to the subfloor or underlayment before you install the carpet squares. Vacuum the surface and wipe it with a damp cloth.

Snap chalk lines between diagonally opposite corners to find the center point for the room. In rooms with unusual shapes, determine the visual center and mark it. Next, snap chalk lines across the center and perpendicular to the walls. This set of guidelines will show you where to start.

(continued)

Lay a base row of carpet squares on each side of the two guidelines. When you reach the walls, make note of how much you will need to cut. You should have the same amount to cut on each side. If not, adjust the center point and realign the squares.

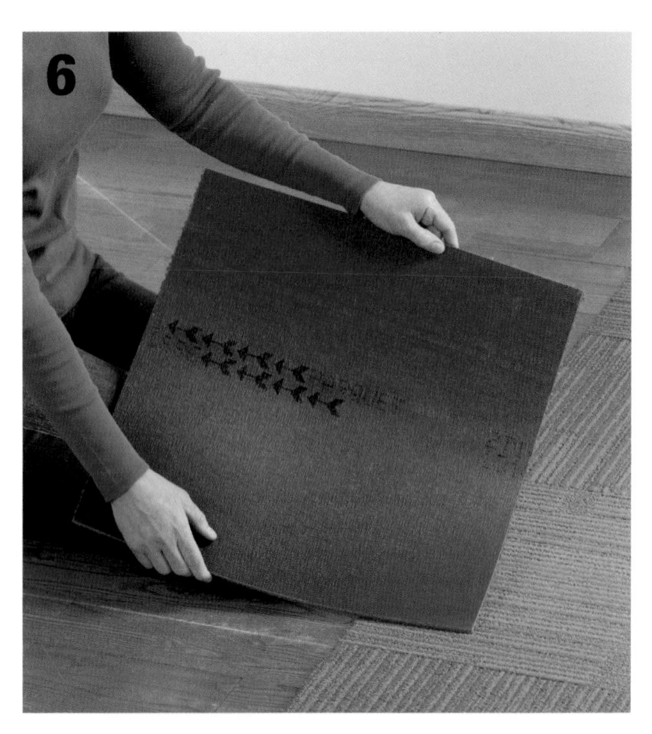

Check the backs of the squares before you apply any adhesive. They should indicate a direction, using arrows or other marks, so that the finished pile has a consistent appearance. If you plan to mix colors, this is the time to establish your pattern.

Fasten the base rows in place using the manufacturer's recommended adhesive. This installation calls for two adhesive dots per square. As you place each square, make sure it is aligned with the guidelines and fits tightly against the next square.

When you reach a wall, flip the last square over. Push it against the wall until it is snug. If you are planning a continuous pattern, align the arrows with the existing squares. If you are creating a parquet pattern, turn the new square 90 degrees before marking it.

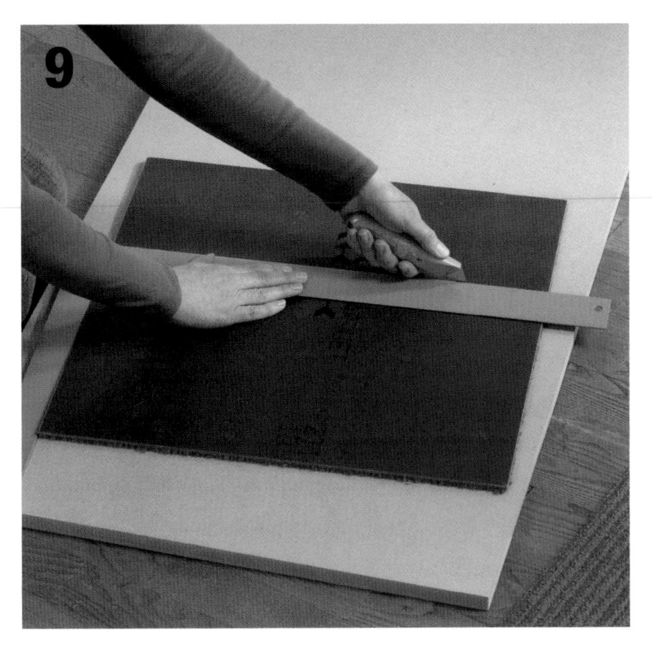

Mark notches or draw a line across the back where the new square overlaps the next-to-last one. Using a sharp carpet knife, a carpenter's square, and a tough work surface, cut along this line. The cut square should fit neatly in the remaining space.

At a door jamb, place a square face up where it will go. Lean the square against the jamb and mark the point where they meet. Move the square to find the other cutline, and mark that as well. Flip the square over, mark the two lines using a carpenter's square, and cut out the corner.

Finish all four base rows before you fill in the rest of the room. As you work, check the alignment of each row. If you notice a row going out of line, find the point where the direction changed, then remove squares back to that point and start again.

Work outward from the center so that you have a known reference for keeping rows straight. Save the cut pieces from the ends. They may be useful for patching odd spaces around doorways, heat registers, radiator pipes, and when you reach the corners.

Finishes & Surface Treatments

One of the most desirable features of hardwood flooring is that it's a natural product. The grain patterns are interesting to the eye and the combination of colors gives any room a soft, inviting glow. The resilience of wood fibers makes a hardwood floor extremely durable, but they are susceptible to changes caused by moisture and aging.

Typically, the first thing to wear out on a hardwood floor is the finish. Refinishing the floor by sanding it with a rented drum sander and applying a topcoat, such as polyurethane, will make your old floor look new. If you don't want to sand the floor, but want to retain the floor's aged glow, or if the boards have been sanded before and are less than 3/8" thick, consider stripping the floor.

Once your floor is finished, you may want to dress it up with a favorite design, border, or pattern. If the wood will not look good refinished, consider painting it. Pages 216 to 223 offer exciting ways to customize your wood floor.

In this Chapter:

- Refinishing Supplies & Techniques
- Stains & Finishes
- Painting Wood Floors
- Creating Faux Stone Floor Tile Designs
- Painting Decorative Borders
- Creating Nature Prints
- Sealing Concrete Floors
- Painting Concrete Floors

Refinishing Supplies & Techniques

Specialty tools and products are necessary for resurfacing or refinishing wood floors. If several scratches, gouges, and stains have damaged the floor, it may be a good idea to resurface it by sanding, using a drum sander (A) for the main floor area and an edger sander (B) for areas next to baseboards. Both tools can be rented from home improvement or rental centers. As a general rule, use the finest-grit sandpaper that's effective for the job. Be sure to get complete operating and safety instructions when renting these tools.

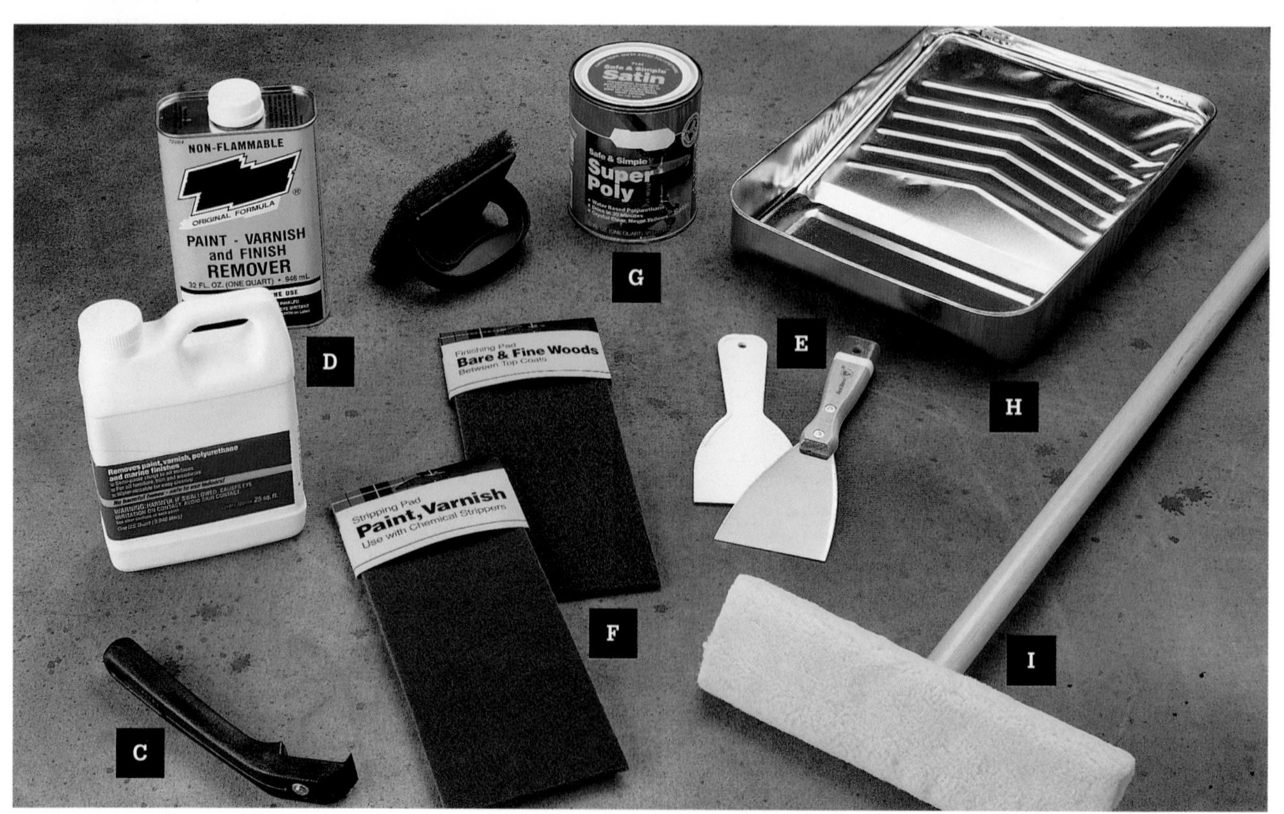

Other products and tools for resurfacing and refinishing floors: Paint scrapers (C) are helpful for removing old finish in corners and other areas that cannot be reached by sanders. When refinishing floors, chemical stripping products (D) are often a more efficient method that yields better results. This is especially true for floors that are uneven, or for parquet and veneered floors, which cannot be sanded. Stripper knives (E) and abrasive pads (F) are used with the stripping products. For the final finish, water-based polyurethane (G) is poured into a paint tray (H) and applied using a wide painting pad with a pole extension (I).

How to Refinish Wood Floors

Staple plastic on all doorways. Place a zip door over the entryway you plan to use for the duration of the project. Use painter's tape and plastic to cover heating and cooling registers, ceiling fans, and light fixtures. Finally, place a fan in a nearby window to blow the circulating dust outside.

Wedge a prybar between the shoe molding and baseboards. Move along the wall as nails loosen. Once removed, place a scrap 2 × 4 board against the wall and, with a pry bar, pry out baseboard at nails (inset). Maintain the gap with wood shims. Drive protruding nails in floor ⅛" below the surface with a nail set.

Practice with the drum sander turned off. Move forward and backward; tilt or raise it off the floor a couple of times. A drum sander is difficult to maneuver. Once it touches the floor it walks forward; if you stop it, it gouges the floor.

For the initial pass with the drum sander, sand with the grain, using 40- or 60-grit sandpaper. For large scratches use 20 or 30. Start two-thirds down the room length on the right side; work your way to the left. Raise drum. Start motor. Slowly lower drum to the floor. Lift the sander off the floor as you approach the wall. Move to the left 2 to 4" and then walk it backwards the same distance you just walked forward. Repeat.

(continued)

When you get to the far left side of the room, turn the machine around and repeat the process. Overlap the sanded two-thirds to feather out the ridgeline. Repeat this drum sanding process 3 or 4 times using 120-grit paper. For the final pass or two use finer paper (150-grit).

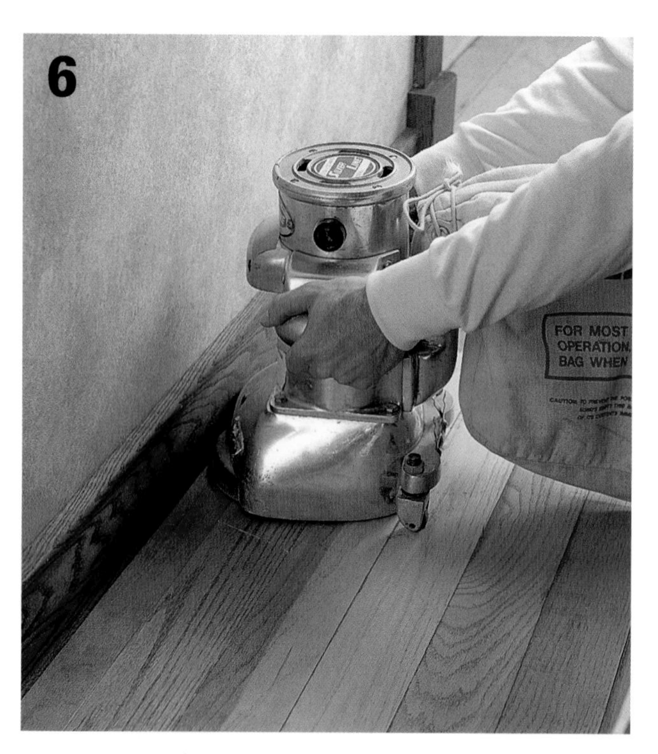

Use a power edge sander along the walls, using the same grit that you last used with the drum sander. Make a succession of overlapping half-circles as you move along the entire perimeter of the room. Next, run a rotary buffer over the floor twice: first with an 80-grit screen and then with a 100-grit screen. Finally, use a random orbital sander to smooth out the floor.

Use a paint scraper to get to corners and hard-to-reach nooks. Pull the scraper toward you with a steady downward pressure. Pull with the grain. Next, sand with a sanding block.

Remove all plastic on doors, windows, and fixtures and then sweep and vacuum to prepare the room for finish. Wipe up fine particles with a tack cloth.

Seal the sanded wood with a 1:1 mixture of water-based polyurethane and water, using a painting pad and pole.

Allow the floor to dry. Buff the surface lightly to remove any raised wood grain, using a medium abrasive pad. Vacuum the surface, using a bristle attachment, then wipe with a tack cloth.

Apply at least two coats of undiluted polyurethane finish to get a hard, durable finish. Allow the finish to dry; repeat step 10 and then add a final coat. Do not overbrush.

When the final coat is dry, buff the surface with water and fine abrasive pad. Wait at least 72 hours before replacing the shoe molding.

How to Chemically Strip a Hardwood Floor

Wearing a respirator and rubber gloves, apply the stripper with a paintbrush. Cover only an area small enough to be scraped within the working time of the stripper.

Scrape off the sludge of stripper and old finish, using a nylon stripper knife. Move the scraper with the wood grain, and deposit the sludge onto old newspapers. After the entire floor is stripped, scrub it with an abrasive pad dipped in a rinsing solvent, such as mineral spirits, that's compatible with your stripping product. Do not use water.

Clean residual sludge and dirt from the joints between floorboards, using a palette knife or putty knife.

Remove stains and discoloration by carefully sanding only the affected area. Use oxalic acid on deep stains.

Touch up sanded areas with wood stain. Test the stain before applying.

Stains & Finishes

Stains are applied to surfaces of unfinished wood floors to change the color and are available in a variety of natural wood tones. Colored stains can be applied to previously stained and finished floors for a colorwashed effect. Consider a colored stain, such as green for a rustic decorating scheme, or white for a contemporary look.

You can stain wood by colorwashing it with diluted latex paint. The colorwash solution will be considerably lighter in color than the original paint color. Use 1 part latex paint and 4 parts water to make a colorwash solution, and experiment with small amounts of paint until you achieve the desired color. Apply the stain or colorwash solution in an inconspicuous area, such as a closet, to test the application method and color before staining the entire floor surface.

Aged finishes (see page 208) give floors timeworn character, making them especially suitable for a rustic or transitional decorating style. Although they appear distressed and fragile, these finishes are actually very durable. Aged finishes are especially suitable on previously painted or stained floors, but they may also be applied to new or resurfaced wood flooring. Up to three coats of paint in different colors may be used.

Look for a water-based stain that's formulated for easy application without lap marks or streaking. Conditioners can help prevent streaking and control grain raise when you're using water-based wood stains. Use a wood conditioner on the wood prior to staining, if recommended by the manufacturer.

Tools & Materials ▸

- Synthetic brush
- Sponge applicators
- Cotton lint-free rags
- Rubber gloves
- Paint pad and pole extension
- Power sander
- Fine- and medium-grit sandpaper
- Vacuum
- Wood conditioner
- Tack cloth
- Water-based stain or latex paint
- High-gloss and satin clear finishes
- Latex enamel paints
- Paint roller
- Hammer
- Chisel
- Awl

Wood Stain Variations

Dark wood tones work well for traditional rooms. White colorwashing over a previously dark stained floor mellows the formal appearance.

Medium, warm wood tones have a casual appearance. White colorwashing over a medium wood tone creates an antiqued look.

Pale neutral stains often are used for contemporary rooms. A blue colorwash can give a pale floor a bold new character.

How to Apply a Stained Finish to a Bare Wood Floor

Sand the floor surface, using fine-grit sandpaper, sanding in the direction of the wood grain. Remove the sanding dust with a vacuum, then wipe the floor with a tack cloth.

Wear rubber gloves when working with any stain product. Stir the stain or colorwash solution thoroughly. Apply the stain or solution to the floor, using a synthetic brush or sponge applicator. Work on one small section at a time. Keep a wet edge and avoid overlapping the brush strokes.

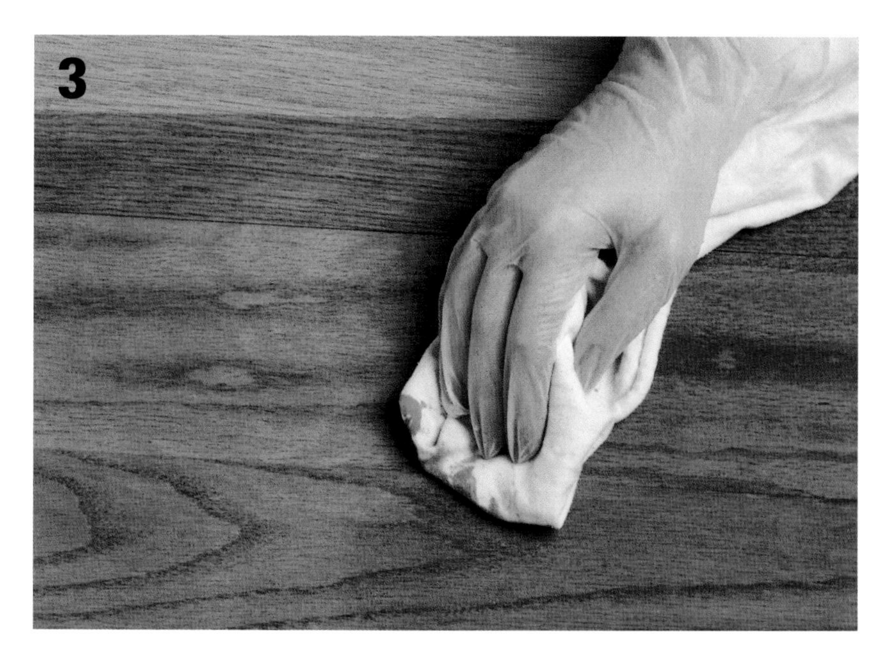

Wipe away excess stain immediately, or after the waiting time recommended by the manufacturer, using a dry, lint-free rag. Wipe across the grain of the wood first, then wipe with the grain. Continue applying and wiping stain until the entire floor is finished. Allow the stain to dry. Sand the floor lightly, using fine-grit sandpaper, then remove any sanding dust with a tack cloth. For a deeper color, apply a second coat of stain and allow it to dry thoroughly.

Apply a coat of high-gloss clear finish to the stained floor, using a sponge applicator or a paint pad with pole extension. Allow the finish to dry. Sand the floor lightly, using fine-grit sandpaper, then wipe with a tack cloth. Apply two coats of satin clear finish following manufacturer's directions.

How to Apply an Aged & Distressed Finish

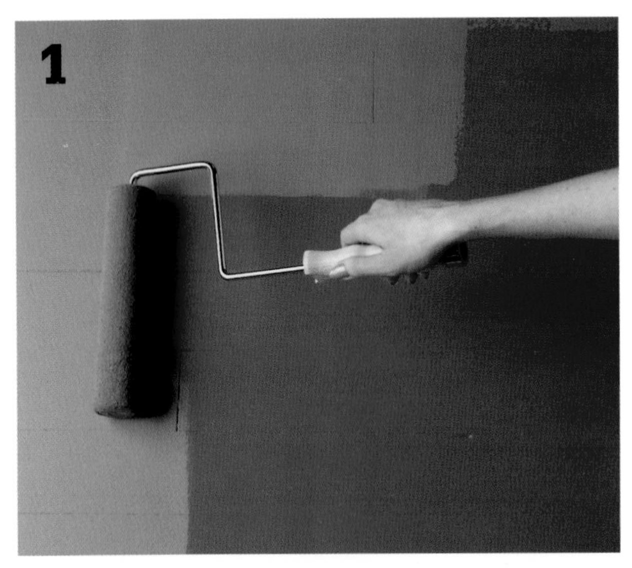

Finish the floor with a painted or stained base coat. Sand the floor lightly, using fine-grit sandpaper. Vacuum the floor and wipe away dust with a tack cloth. Apply two or three coats of enamel, using a different color of paint for each coat. Allow the floor to dry between coats. Sand the floor lightly between coats, using fine-grit sandpaper, and wipe away dust with a tack cloth.

Sand the floor surface with medium-grit sandpaper, sanding harder in some areas to remove the top and middle coats of paint, using a power sander. Avoid sanding beyond the base coat of paint or stain.

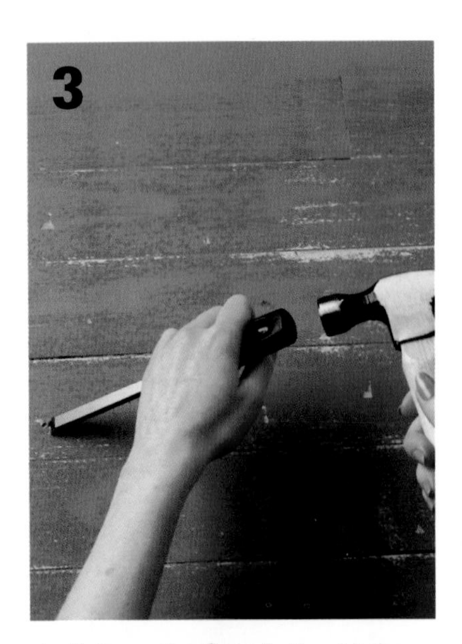

To distress the floor further, hit the wood with the head of a hammer or a chain. Gouge the boards with a chisel, or pound holes randomly, using an awl. Create as many imperfections as desired, then sand the floor lightly with fine-grit sandpaper. Apply two coats of satin clear finish, allowing the floor to dry completely between coats.

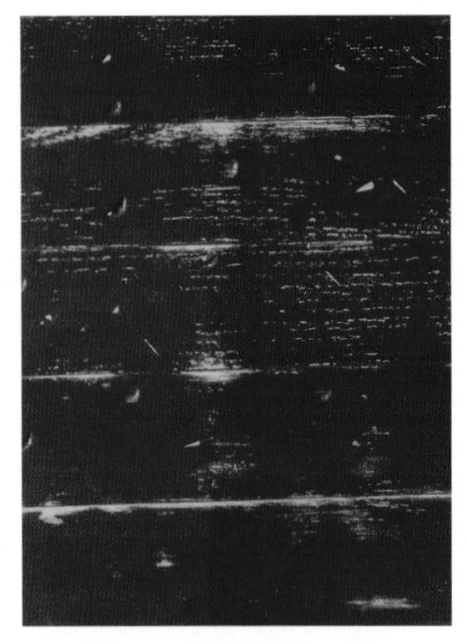

Variation: Two coats of dark green paint were applied over a previously stained floor. Sanding revealed the stain in some areas. The floor was further distressed, using a hammer, chisel, and awl.

Variation: Maroon base coat and light rose top coat were painted over a previously stained floor. Sanding created an aged look suitable for a cottage bedroom.

Painting Wood Floors

Paint is a quick, cost-effective way to cover up wood floors that may need work, but a floor doesn't have to be distressed or damaged to benefit from paint. Floors in perfect condition in both formal and informal spaces can be decorated with paint to add color and personality. For example, one could unify a space by extending a painted floor through a hallway to a staircase. And stencil designs or faux finishes can make an oversized room feel cozy and inviting. There are even techniques for disguising worn spots.

Tools & Materials ▸

- Lacquer thinner
- Primer
- Latex paint specifically for floors
- Wood putty
- Wide painter's tape
- Drop cloths
- Paint roller and tray
- 4"-wide paintbrush
- Extension pole
- Paint scraper
- Hammer
- Pole sander
- Sandpaper
- Putty knife
- Nail set
- Dusk mask

To paint a wood floor you must apply primer first, then apply the paint, and follow that with a polyurethane sealer. Make sure the products you choose are specifically for floors.

Primer is a specially-formulated paint that is used to seal raw surfaces and to provide a base coat of paint that succeeding finish coats can better adhere to. Water-based polyurethane is a clear finish used for coating natural or stained wood. It provides a durable, glossy surface that is highly resistant to water and allows for easy clean-up of spills. A sealer is a finish coating, either clear or pigmented, that is applied on top of the paint.

How to Paint Wood Floors

Prepare the painting area by first moving furniture. Lift pieces instead of dragging them to prevent gouges. Sweep or vacuum.

Use a paint scraper to remove smooth rough spots. Use a pole sander to sand with the grain of the wood. For coarse wood, use medium-grit sandpaper. Scuff glossy hardwoods with fine sandpaper (120-grit) for good adhesion.

When finished sanding, sweep or vacuum. Use a damp cloth to remove fine dust. Use a cloth dampened with lacquer thinner for a final cleaning. If you see any nails sticking up, tap them down with a hammer and nail set.

Protect the baseboards with wide painter's tape. Press the tape edges down so paint doesn't seep underneath.

(continued)

Mix primer well (see Step 6 for mixing technique). Use a 4"-wide brush to apply the primer around the perimeter of the room. Then paint the remaining floor with a roller on an extension pole. Allow the primer to dry.

To mix paint, pour half of the paint into another can, stirring the paint in both containers with a wooden stir stick before recombining them. As you stir, you want a smooth consistency.

Paint. Use a 4" brush to apply a semi-gloss paint around the border. To paint the rest of the floor, use a roller on an extension pole. Always roll from a dry area to a wet area to minimize lap lines. Allow paint to dry. Apply second coat of paint. Allow to dry.

Apply 2 or 3 coats of a matte-finish, waterbased polyurethane sealer, using a painting pad on a pole. Allow the paint to dry. Sand with a pole sander, using fine sandpaper. Clean up dust with a tack cloth.

How to Paint a Checkerboard Pattern

If your wood floor is in poor condition, it can be camouflaged with a design, such as a classic checkerboard pattern. Proper preparation is essential for lasting results. If you've already painted the floor based on the instructions from the first half of this project, you are well on your way and you have the base color already painted. You just have to paint in the darker colored squares (skip to Step 3).

Remove the finish of a previously stained and sealed wood floor by sanding the surface with sandpaper on a pole, using fine-grit sandpaper. Alternatively, you may use a power sander with fine-grit sandpaper. Vacuum the entire floor, then wipe it with a tack cloth to remove all sanding dust.

Use painter's tape to cover the baseboards. Paint the entire floor with the lighter of the two paint colors. Allow the paint to dry thoroughly.

(continued)

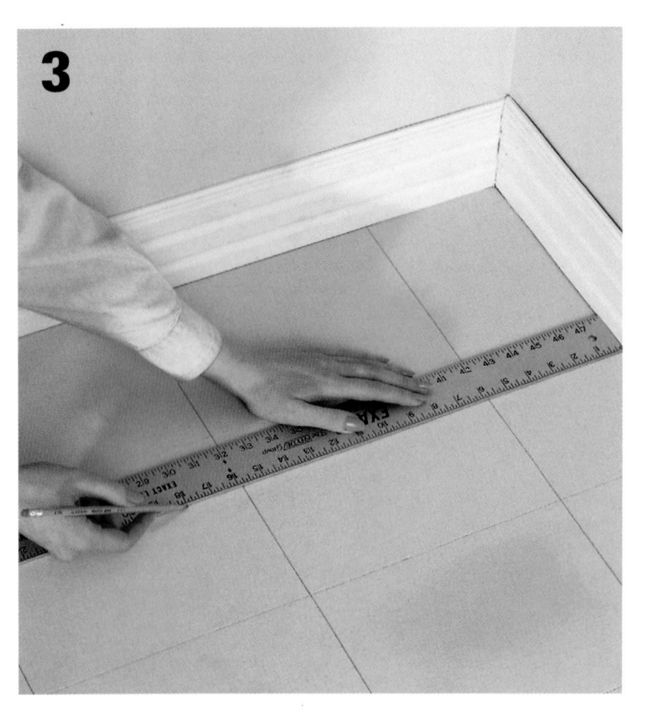

Measure the entire floor. Now, determine the size of squares you'll use. Plan the design so the areas of the floor with the highest visibility, such as the main entrance, have full squares. Place partial squares along the walls in less conspicuous areas. Mark the design lines on the floor, using a straightedge and pencil.

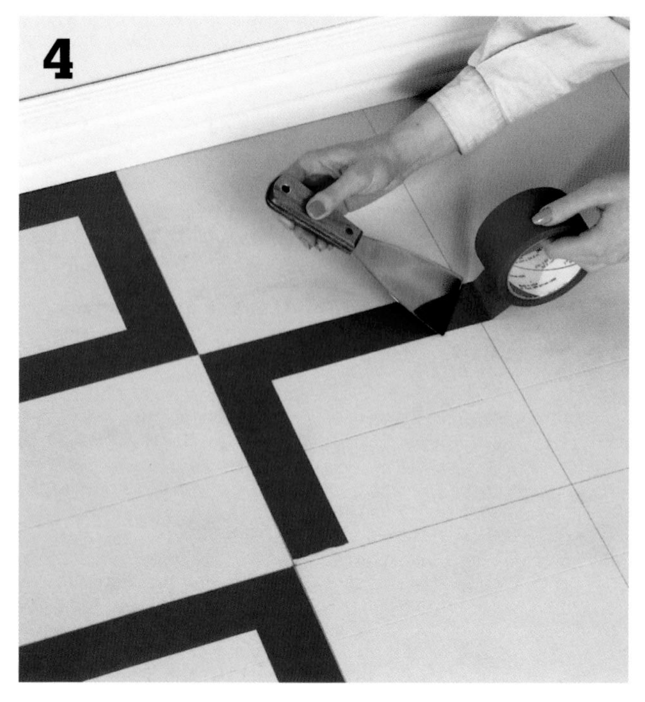

Using painter's tape, outline the squares that are to remain light in color. Press firmly along all edges of the tape, using a putty knife, to create a tight seal.

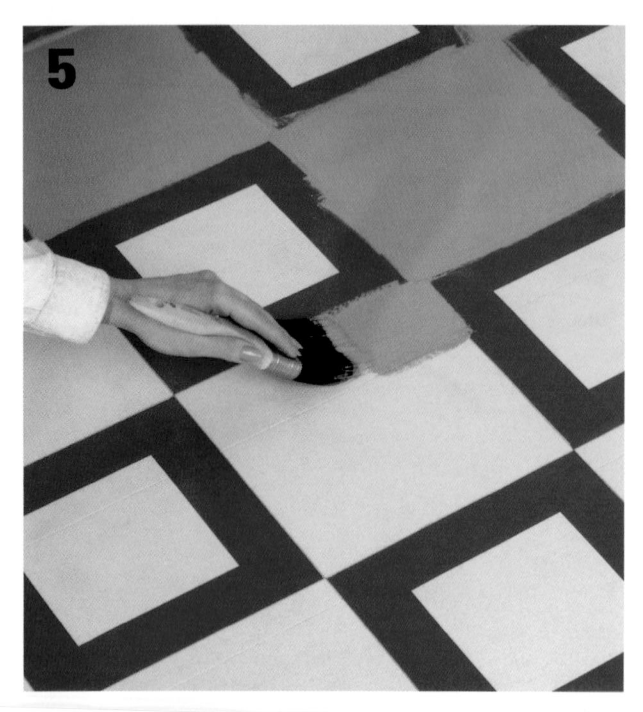

Paint the remaining squares with the dark paint color. Paint in small areas at a time. Once you have painted the entire box and a few surrounding boxes, remove the masking tape from the painted squares. Be sure to remove the tape before the paint completely dries.

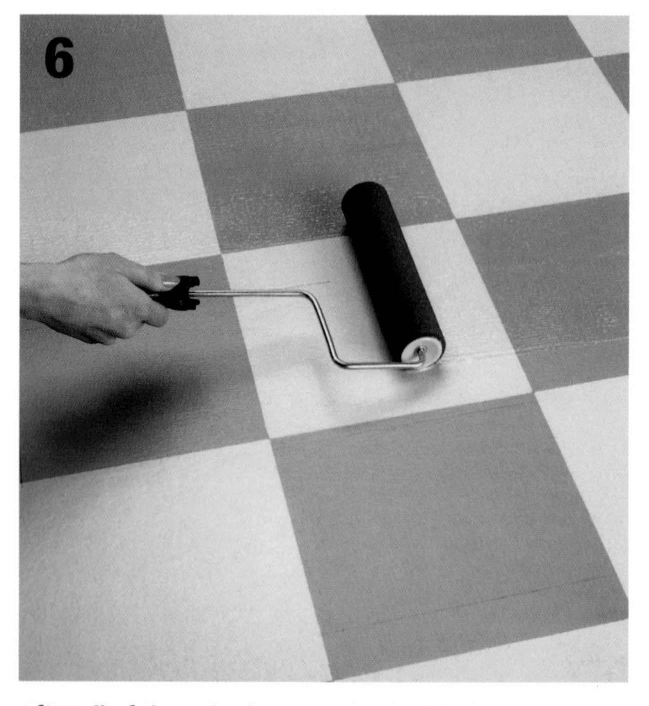

After all of the paint has completely dried, apply a coat of high-gloss clear finish, using a paint roller or paint pad with a pole extension. Allow the finish to dry.

How to Make a Painted Floor Cloth

Trim selvages from the canvas. Mark the canvas to the desired size, using a pencil, carpenter's square, and straightedge. Cut the canvas to size.

Machine stitch around the canvas $\frac{1}{4}$" from raw edges. Stitch a second row of stitching $\frac{1}{8}$" from raw edges. Press the canvas so it lies flat.

Place the canvas on a plastic drop cloth. Using a paint roller, apply the background paint color, taking care not to crease the canvas. Roll the paint in all directions to penetrate the fabric. Allow it to dry. Apply additional coats of paint as necessary, allowing the paint to dry overnight. Trim any loose threads.

Mark your design, using a pencil. Paint the design, applying one color at a time. Use a fine-pointed brush for outlining and a wider brush for filling in the design areas. Allow the paint to dry 24 hours.

Apply sealer, using a synthetic-bristle paintbrush. Allow the sealer to dry for several hours. Apply two additional coats of sealer, following the manufacturer's instructions for drying time.

Place the canvas on the floor, making sure it lies perfectly flat.

Creating Faux Stone Tile Designs

A painted finish that mimics unpolished stone can be applied to a floor using a stippler or pieces of newspaper. The stippling technique results in a relatively smooth textured finish with blended colors, while the newspaper method creates an unpolished stone finish with depth, color variation, and a rough visual texture. Adding rustic grout lines creates the look of expensive stone tiles.

A variety of earth-tone glazes can be combined to make a stone finish in the desired color. It's best to limit your selection to two or three colors.

To create grout lines, apply masking tape to the surface in a grid pattern before applying the faux finish. Once the finish has been applied and is dry, remove the tape and paint the lines with a rounded artist's brush. You can get a rustic look by painting the lines by hand or create the illusion of depth by "shadowing" the grout lines with a marker that's darker than the base coat color.

For faux stone flat glazes, mix one part latex paint, one part latex paint conditioner, and one part water for each shade of earth-toned flat glaze. For faux stone washes, mix one part flat latex paint and two to three parts water for each shade of wash. Dilute the paint with water until it reaches the consistency of ink.

Tools & Materials ▸

- Low-napped roller
- Sponge applicator or paintbrush
- Stippler or newspaper
- White low-luster latex enamel paint
- Flat latex paint in earth-tone shades
- Latex paint conditioner
- Cheesecloth
- Flat latex paint in white and earth-tone
- Matte clear finish or aerosol
- Matte clear acrylic sealer (optional)
- Latex or craft acrylic paint to contrast with stone finish

How to Apply a Faux Stone Finish Using the Stippling Method

Apply a base coat of white low-luster latex enamel to the surface, using an applicator suitable to the surface size. Allow the paint to dry. Mask off grout lines, if desired.

Apply a flat earth-tone glaze in random strokes, using a sponge applicator or paintbrush. Cover about half of the surface. Repeat the random strokes with another color glaze in the remaining areas, leaving some small areas of the base coat unglazed.

Using a stippler, lightly brush over the surface. Blend the colors as desired, leaving some areas very dark and others light enough for the base coat to show through. Add white and black glazes, or earth-tone glazes, as desired. Blend the colors with the stippler, and then allow the paint to dry.

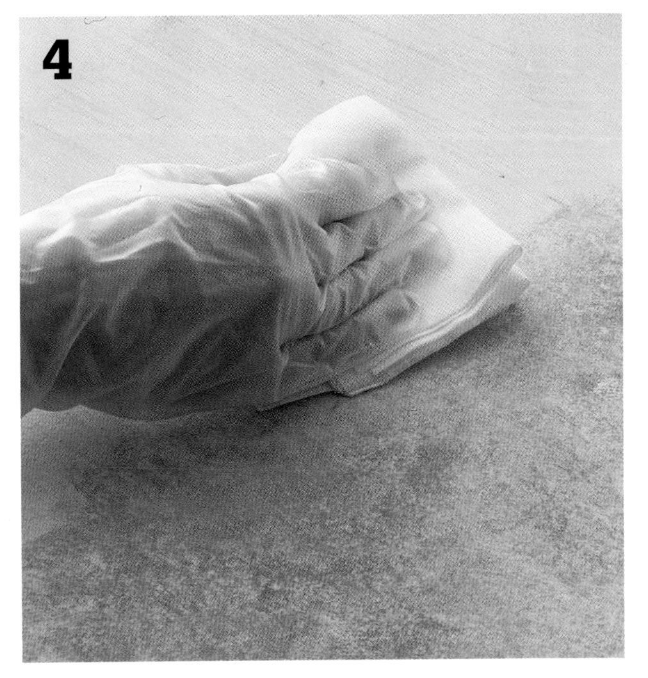

Apply a white wash to the entire surface. Dab with a wadded cheesecloth to soften the finish. Allow the paint to dry, then apply matte clear finish or matte aerosol clear acrylic sealer, if desired.

How to Apply a Faux Stone Finish Using the Newspaper Method

Follow steps 1 and 2 on page 217. Apply a white wash in desired areas, and apply an earth-tone wash in other areas.

Fold a sheet of newspaper to several layers. Lay it flat over one area of the surface and press it into the glaze. Lift the paper, removing some of the glaze. Repeat in other areas, using the same newspaper, turning it in different directions to blend the colors roughly.

Add more color to an area by spreading glaze onto the newspaper and laying it flat on the surface. Repeat as necessary until the desired effect is achieved. Leave some dark accent areas in the finish as well as an occasional light spot. Use the same newspaper throughout. Allow the paint to dry.

Apply a white wash to the entire surface. Dab with a wadded cheesecloth to soften the colors. Allow the paint to dry, then apply a matte clear finish or matte aerosol clear acrylic sealer, if desired.

How to Apply Rustic Grout Lines

Apply a base coat of white low-luster latex enamel to the surface, using an applicator suitable to the surface size. Allow the paint to dry. Plan the placement of the grout lines, and mark the points of intersection, using a pencil and straightedge.

Stretch ¼" masking tape taut, and apply it to the surface in horizontal lines, positioning the tape lines just under the marked points. Repeat this process for the vertical lines, positioning the tape just right of the marks. Press the tape firmly in place with your fingers, but don't burnish the tape.

Apply the desired faux finish (page 213). Allow the finish to dry, then carefully remove the tape lines.

Paint over the grout lines freehand, using a round artist's brush and a grout line glaze in a color that contrasts pleasingly with the faux finish. Allow the lines to have some irregularity in thickness and density. Allow the paint to dry. Apply a finish or sealer to the entire surface, as desired.

Painting Decorative Borders

A painted striped border with a block print design is a great way to add a personal touch to your floor. The stripes and designs effectively frame an attractive wood floor. The actual painting is the easy part of the project. The more time-consuming aspect is laying out the design on the floor and taping the paint lines. Take your time applying the tape to make sure it's pressed firmly to the floor, or paint could run under it.

Closed-cell foam for the block-print design is available as thin, pressure-sensitive sheets, neoprene weather-stripping tape, neoprene sheets, and computer mouse pads.

Tools & Materials ▸

- Fine-grit sandpaper
- Painter's masking tape
- Putty knife
- Straightedge
- Tape measure
- Vacuum
- Tack cloth
- Paintbrush
- Paint
- Closed-cell foam
- Wood block
- Acrylic paint extender
- Felt
- Glass or acrylic sheet
- High-gloss clear finish
- Sponge applicator
- Satin clear finish

How to Paint Decorative Borders

Sand the surface of a previously stained and sealed wood floor lightly in the area to be painted, using fine sandpaper, to degloss the finish and improve paint adhesion. Vacuum the entire floor, and wipe with tack cloth.

Measure and mark the design lines for the border on the floor, using a straightedge and pencil. Mask off stripes in the design, using painter's masking tape. Press the tape firmly along edges with a putty knife to prevent paint from seeping under the tape.

Apply paint for the stripes, using a paintbrush. Remove the masking tape. Allow the paint to dry.

(continued)

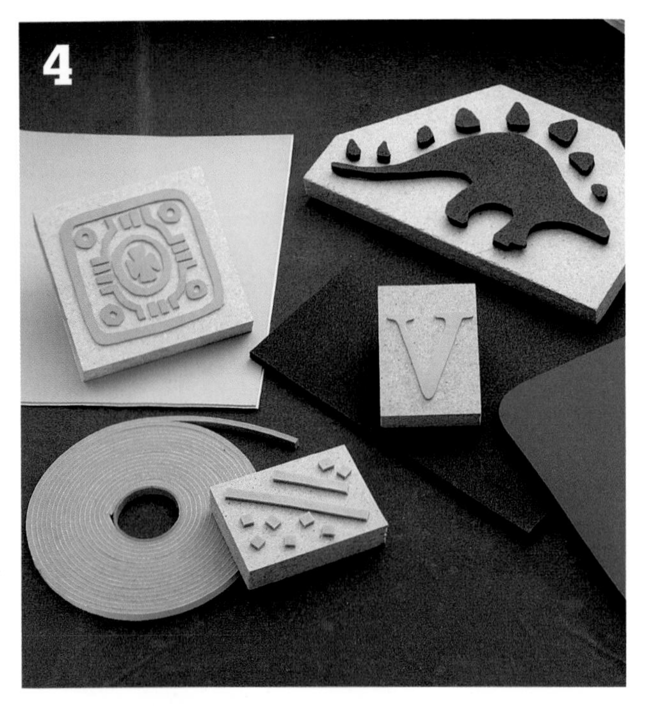

Make printing blocks from closed-cell foam that's cut to desired shapes and attached to a wood block.

Thin the paint slightly with an acrylic paint extender, about three to four parts paint to one part extender. Cut a piece of felt, larger than the printing block. Place the felt pad on glass or acrylic sheet. Pour the paint mixture onto the felt, allowing the paint to saturate the pad.

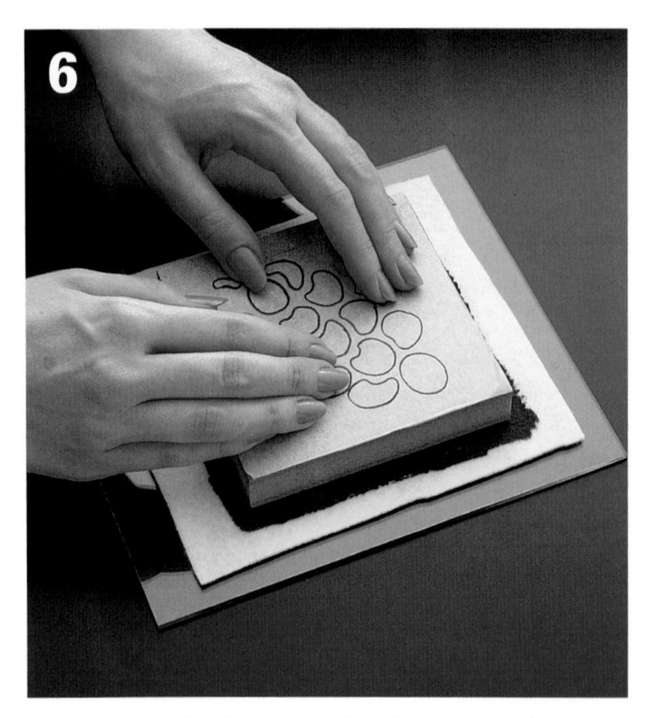

Press the printing block onto the felt pad, coating the surface of the foam evenly with paint.

Press the printing block onto the floor, applying firm, even pressure to the back of the block. Remove the block by pulling it straight back off the floor to avoid smudging. Apply a coat of high-gloss clear finish over the entire floor, using a sponge applicator. Allow the finish to dry, then sand lightly with fine-grit sandpaper. Wipe with tack cloth. Apply two coats of satin clear finish.

Creating Nature Prints

Use leaves to create a unique imprint on your floor. Leaves can be collected from your own backyard or found at a local florist. Experiment with the printing process on paper to determine which leaves provide the desired finished result. Printing with the back side of the leaf may provide more detail in the finished print.

Nature print designs can be applied easily over previously stained or color-washed and finished floors. A wood floor can be embellished with corner designs or a border. For a parquet floor, apply leaf prints randomly to the centers of the wood squares.

Tools & Materials ▸

- Fine-grit sandpaper
- Tack cloth
- Synthetic brush
- Sponge applicator
- Wax paper
- Rubber brayer (about 4" wide)
- Lint-free cloth
- Craft acrylic paints
- High-gloss and satin clear finishes
- Fresh leaves

How to Print Floor Designs

Press the leaves flat by placing them between pages of a large book for about an hour. Sand the surface of a previously stained and sealed wood floor lightly in the area to be printed, using fine-grit sandpaper. Wipe with a tack cloth. Apply a thin layer of craft acrylic paint to the back side of the leaf, using a sponge applicator.

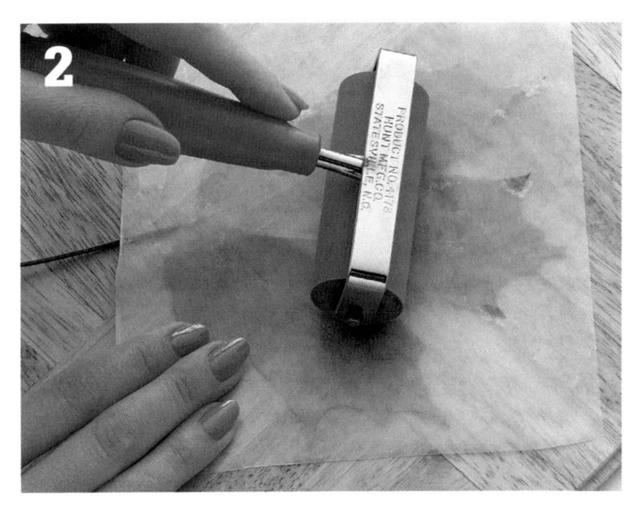

Position the leaf, paint-side down, over the floor in the desired location of the print, then cover it with wax paper. Roll the brayer over the leaf to make an imprint. Carefully remove the wax paper and leaf. Remove any unwanted paint lines from the imprint, using a damp cloth before the paint dries. Allow the paint to dry. Repeat the printing process for a desired number of leaf prints. Apply two coats of clear finish over the designs.

Sealing Concrete Floors

Most people are accustomed to thinking of concrete primarily as a utilitarian substance, but it can also mimic a variety of flooring types and be a colorful and beautiful addition to any room. Whether your concrete floor is a practical surface for the garage or an artistic statement of personal style in your dining room, it should be sealed.

Concrete is a hard and durable building material, but it is also porous—so it is susceptible to staining. Many stains can be removed with the proper cleaner, but sealing and painting prevents oil, grease, and other stains from penetrating the surface in the first place; and cleanup is a whole lot easier.

Even after degreasing a concrete floor, residual grease or oils can create serious adhesion problems for coatings of sealant or paint. To check to see whether your floor has been adequately cleaned, pour a glass of water on the concrete floor. If it is ready for sealing, the water will soak into the surface quickly and evenly. If the water beads, you may have to clean it again. Detergent used in combination with a steam cleaner can remove stubborn stains better than a cleaner alone.

There are four important reasons to seal your concrete floor: to protect the floor from dirt, oil, grease, chemicals, and stains; to dust-proof the surface; to protect the floor from abrasion and sunlight exposure; and to repel water and protect the floor from freeze-thaw damage.

Tools & Materials ▸

- Acid-tolerant pump sprayer
- Alkaline-base neutralizer
- Sealant
- Rubber Boots
- Garden hose with nozzle
- Acid-tolerant bucket
- Eye protection
- Paint roller frame
- Soft-woven roller cover
- Paint
- Roller tray
- Wet vacuum
- High-pressure washer
- Paintbrush
- Respirator
- Stiff bristle broom
- Extension handle
- Rubber gloves

How to Seal Concrete Floors

Clean and prepare the surface by first sweeping up all debris. Next, remove all surface muck: mud, wax, and grease. Finally, remove existing paints or coatings. See the chapter on cleaning concrete for tips on what to use to remove a variety of common stains.

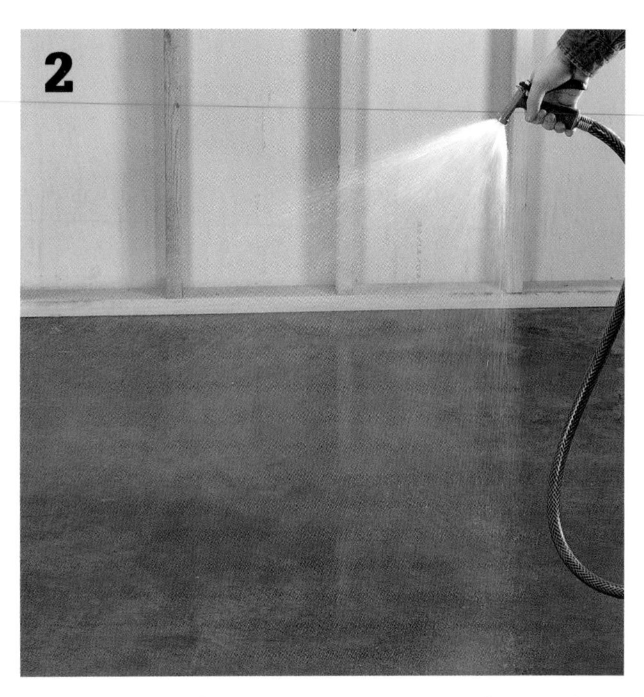

Saturate the surface with clean water. The surface needs to be wet before acid etching. Use this opportunity to check for any areas where water beads up. If water beads on the surface, contaminants still need to be cleaned off with a suitable cleaner or chemical stripper.

Test your acid-tolerant pump sprayer with water to make sure it releases a wide, even mist. Once you have the spray nozzle set, check the manufacturer's instructions for the etching solution and fill the pump sprayer with the recommended amount of water.

Add the acid etching contents to the water in the acid-tolerant pump sprayer (or sprinkling can). Follow the directions (and mixing proportions) specified by the manufacturer. Use caution.

(continued)

Apply the acid solution. Using the sprinkling can or acid-tolerant pump spray unit, evenly apply the diluted acid solution over the concrete floor. Do not allow acid solution to dry at any time during the etching and cleaning process. Etch small areas at a time, 10 × 10 ft. or smaller. If there is a slope, begin on the low side of the slope and work upward.

Use a stiff bristle broom or scrubber to work the acid solution into the concrete. Let the acid sit for 5 to 10 minutes, or as indicated by the manufacturer's directions. A mild foaming action indicates that the product is working. If no bubbling or fizzing occurs, it means there is still grease, oil, or a concrete treatment on the surface that is interfering. If this occurs, follow steps 7 to 12 and then clean again.

Once the fizzing has stopped, the acid has finished reacting with the alkaline concrete surface and formed pH-neutral salts. Neutralize any remaining acid with an alkaline-base solution. Put a gallon of water in a 5-gallon bucket and then stir in an alkaline-base neutralizer. Using a stiff bristle broom, make sure the concrete surface is completely covered with the solution. Continue to sweep until the fizzing stops.

Use a garden hose with a pressure nozzle or, ideally, a pressure washer in conjunction with a stiff bristle broom to thoroughly rinse the concrete surface. Rinse the surface 2 to 3 times. Re-apply the acid (repeat Steps 5, 6, 7, and 8).

If you have any leftover acid you can make it safe for your septic system by mixing more alkaline solution in the 5-gallon bucket and carefully pouring the acid from the spray unit into the bucket until all of the fizzing stops.

Use a wet vacuum to clean up the mess. Some sitting acids and cleaning solutions can harm local vegetation, damage your drainage system, and are just plain environmentally unfriendly. Check your local disposal regulations for proper disposal of the neutralized spent acid.

To check for residue, rub a dark cloth over a small area of concrete. If any white residue appears, continue the rinsing process. Check for residue again.

Let the concrete dry for at least 24 hours and sweep up dust, dirt, and particles leftover from the acid etching process. Your concrete should now have the consistency of 120-grit sandpaper and be able to accept concrete sealants.

(continued)

Once etched, clean, and dry, your concrete is ready for clear sealer or liquid repellent. Mix the sealer in a bucket with a stir stick. Lay painter's tape down for a testing patch. Apply sealer to this area and allow to dry to ensure desired appearance. Concrete sealers tend to make the surface slick when wet. Add an anti-skid additive to aid with traction, especially on stairs.

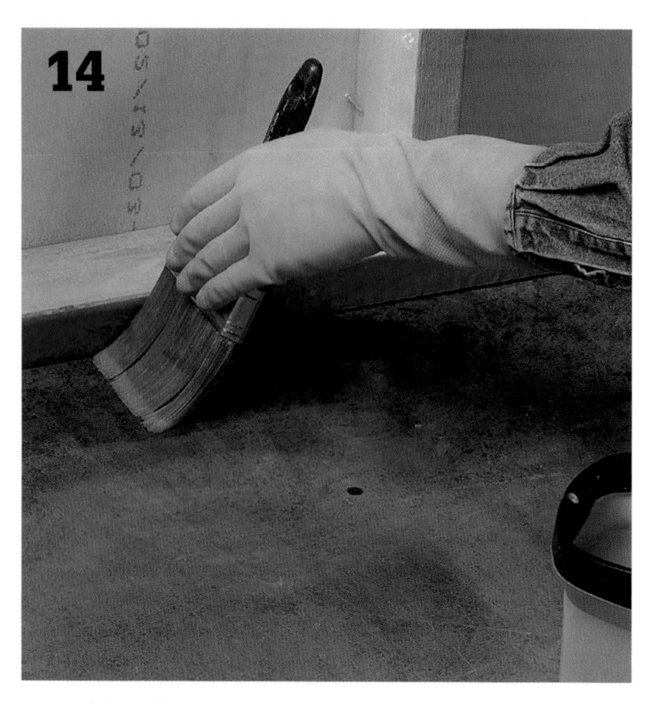

Use wide painter's tape to protect walls and then, using a good quality 4"-wide synthetic bristle paintbrush, coat the perimeter with sealer.

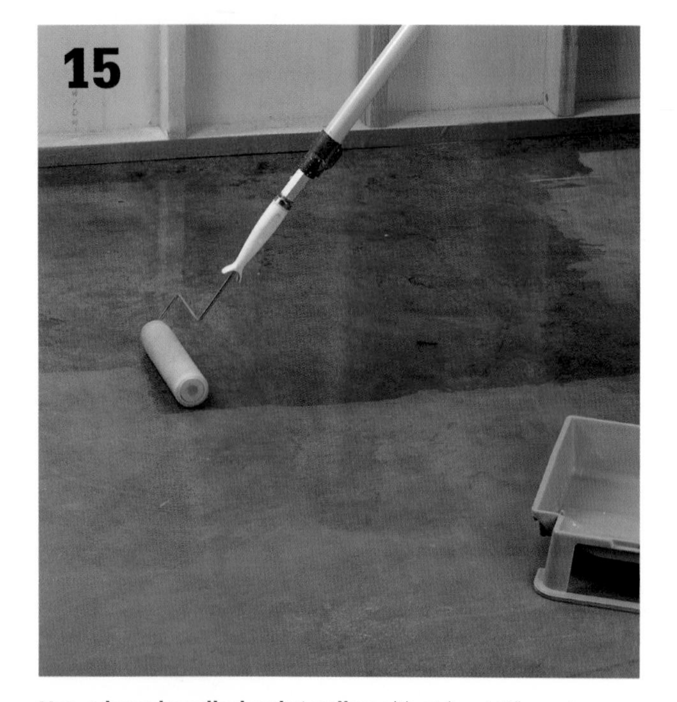

Use a long-handled paint roller with at least ½" nap to apply an even coat to the rest of the surface. Do small sections at a time (about 2' × 3'). Work in one orientation (e.g., north to south). Avoid lap marks by always maintaining a wet edge. Do not work the area once the coating has partially dried; this could cause it to lift from the surface.

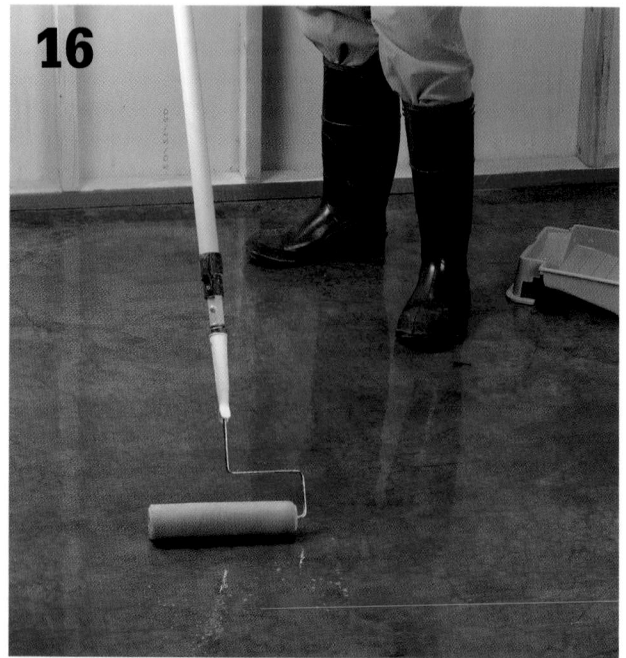

Allow surface to dry according to the manufacturer's instructions, usually 8 to 12 hours minimum. Then apply a second coat in the opposite direction to the first coat, so if the first coat was north to south, the second coat should be east to west.

Painting Concrete Floors

This book includes two ways of building new floors on top of concrete. But sometimes it just isn't practical to add a new layer. Maybe your basement ceiling is already low, and you need to preserve as much headroom as possible. Or maybe you don't use the space often enough to justify a full makeover.

To give concrete flooring a facelift, nothing is simpler than paint. You can protect the surface from dirt and stains with a clear sealer, give it a translucent color wash, or cover it with an opaque floor paint. Just make sure the surface is clean, dry, level, smooth, and free of any grease or wax.

Before you choose a surface treatment for your concrete floor, do some browsing. Try your local library, the web, and building supply and paint stores. Once you find the look you're after, the directions here will help you create it.

Basements seldom have adequate ventilation for working with paint. Set up fans to keep the air moving. If your concrete floor is on a porch or garage, just leave the doors open.

Tools & Materials ▸

- Bleach solution
- Cleaning supplies
- Electric fan(s)
- Eye protection
- Nylon paint brush
- Nylon-bristle brush
- Paint roller and tray
- Patching compound
- Roller extension handle
- Rubber gloves

How to Paint Concrete Floors

Concrete floors can hold paint made for them, but first the concrete must be clean and dry. Sweep, vacuum, and mop the floor thoroughly. To remove any stains, scrub the floor with solution of 1 part bleach to 3 parts water. Wear eye protection and rubber gloves.

Rinse the surface well with clean water and let it dry.

Following the manufacturer's directions, use a concrete patching compound to repair any cracks. Make sure the floor surface does not flake or crumble anywhere.

Test the absorption of the concrete by sprinkling some water on the floor. If the water is absorbed quickly, paint will probably bond well. If it beads up, you should probably use the acid etching method shown on pages 225 to 227. After etching, let the floor dry at least overnight.

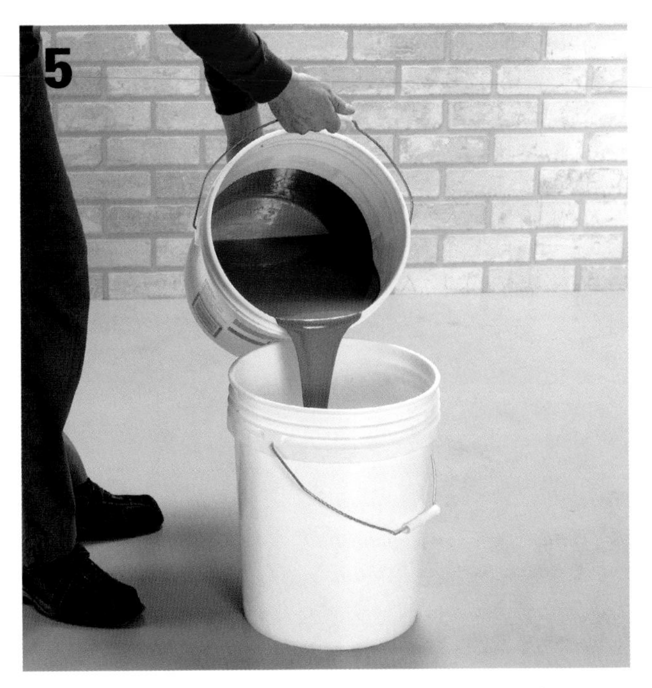

If you expect to use more than one container of paint, open them all and mix them together for a uniform color. You do not need to thin a paint for use on a floor. One exception is if you use a sprayer that requires thinned paint.

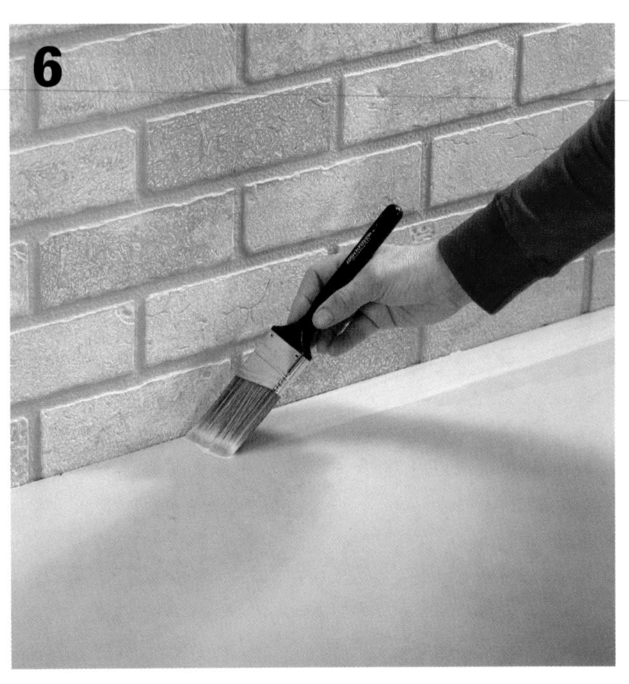

Using a nylon brush, such as a 2½-inch sash brush, cut in the sides and corners with primer. This creates a sharp, clean edge. Start this way for the top coat as well.

Using a roller pad with the nap length recommended by the manufacturer, apply a primer coat to the surface. Start at the corner farthest away from the door, and back up as you work. Allow the primer to dry for at least 8 hours.

With a clean roller pad, apply the first top coat. Make the top coat even but not too thick, then let it dry for 24 hours. If you choose to add another top coat, work the roller in another direction to cover any thin spots. Let the final coat dry another day before you walk on it.

FL. OZ.

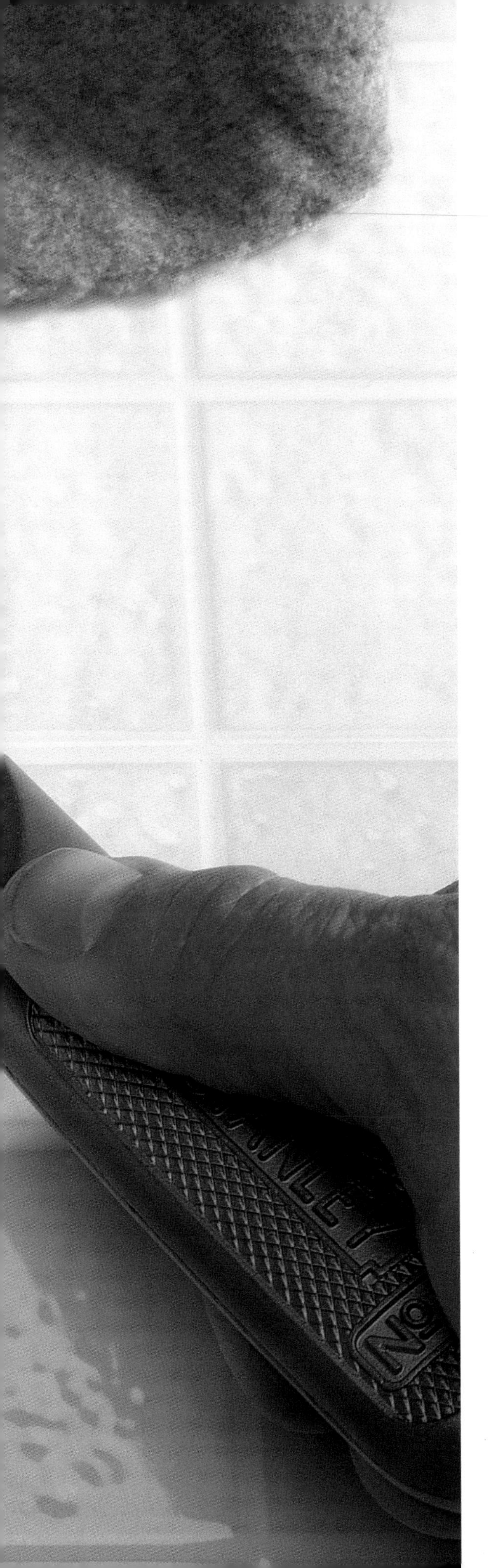

Repairs

Floor coverings wear out faster than other interior surfaces because they get more wear and tear. Surface damage can affect more than just appearance. Scratches in resilient flooring and cracks in grouted tile joints can let moisture into the floor's underpinnings. Hardwood floors lose their finish and become discolored. Loose boards squeak.

Underneath the finished flooring, moisture ruins wood underlayment and the damage is passed on to the subfloor. Bathroom floors suffer the most from moisture problems. Subflooring can pull loose from joists, causing floors to become uneven and springy.

You can fix these problems yourself, such as squeaks, a broken stair tread, damaged baseboard and trim, and minor damage to floor coverings, with the tools and techniques shown on the following pages.

In this Chapter:

- Eliminating Floor Squeaks
- Eliminating Stair Squeaks
- Replacing a Broken Stair Tread
- Replacing Trim Moldings
- Repairing Hardwood
- Repairing Vinyl
- Repairing Carpet
- Repairing Ceramic Tile
- Repairing Concrete

Eliminating Floor Squeaks

Floors squeak when floorboards rub against each other or against the nails securing them to the subfloor. Hardwood floors squeak if they haven't been nailed properly. Normal changes in wood make some squeaking inevitable, although noisy floors sometimes indicate serious structural problems. If an area of a floor is soft or excessively squeaky, inspect the framing and the foundation supporting the floor.

Whenever possible, fix squeaks from underneath the floor. Joists longer than 8 feet should have X-bridging or solid blocking between each pair to help distribute the weight. If these supports aren't present, install them every 6 feet to stiffen and help silence a noisy floor.

Tools & Materials ▸

- Drill
- Hammer
- Nail set
- Putty knife
- Wood screws
- Flooring nails
- Wood putty
- Graphite powder
- Dance-floor wax
- Pipe straps
- Hardwood shims
- Wood glue

How to Eliminate Floor Squeaks

If you can access floor joists from underneath, drive wood screws up through the subfloor to draw hardwood flooring and the subfloor together. Drill pilot holes and make certain the screws aren't long enough to break through the top of the floorboards. Determine the combined thickness of the floor and subfloor by measuring at cutouts for pipes.

When you can't reach the floor from underneath, surface-nail the floor boards to the subfloor with ring-shank flooring nails. Drill pilot holes close to the tongue-side edge of the board and drive the nails at a slight angle to increase their holding power. Whenever possible, nail into studs. Countersink the nails with a nail set and fill the holes with tinted wood putty.

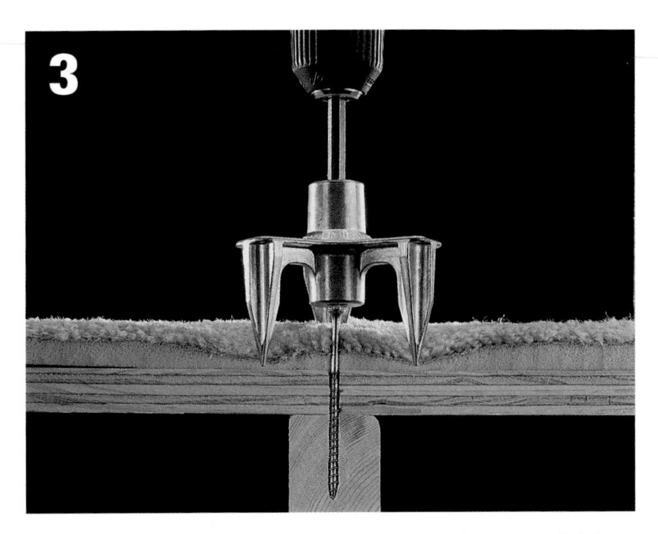

Eliminate squeaks in a carpeted floor by using a special floor fastening device, called a Squeeeeek No More, to drive screws through the subfloor into the joists. The device guides the screw and controls the depth. The screw has a scored shank, so once it's set, you can break the end off just below the surface of the subfloor.

Eliminate squeaks in hardwood floors with graphite powder, talcum powder, powdered soap, mineral oil, or liquid wax. Remove dirt and deposits from joints, using a putty knife. Apply graphite powder, talcum powder, powdered soap, or mineral oil between squeaky boards. Bounce on the boards to work the lubricant into the joints. Clean up excess powder with a damp cloth. Liquid wax is another option, although some floor finishes, such as urethane and varnish, are not compatible with wax, so check with the flooring manufacturer. Use a clean cloth to spread wax over the noisy joints, forcing the wax deep into the joints.

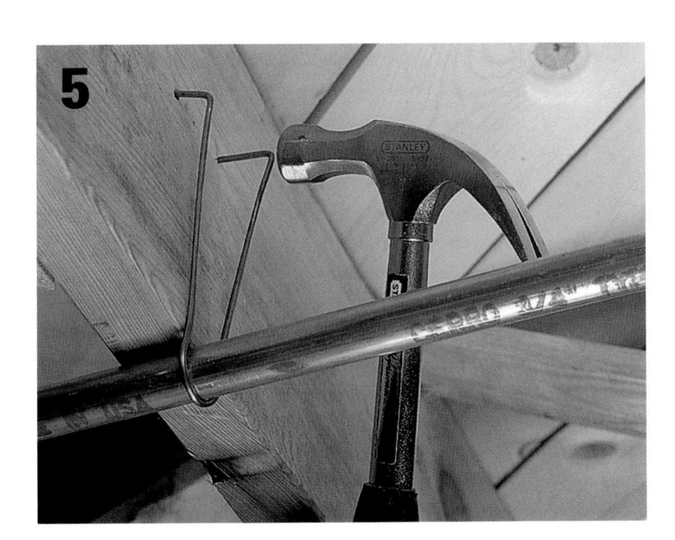

In an unfinished basement or crawl space, copper water pipes are usually hung from floor joists. Listen for pipes rubbing against joists. Loosen or replace wire pipe hangers to silence the noise. Pull the pointed ends of the hanger from the wood, using a hammer or pry bar. Lower the hanger just enough so the pipe isn't touching the joist, making sure the pipe is held firmly so it won't vibrate. Renail the hanger, driving the pointed end straight into the wood.

The boards or sheeting of a subfloor can separate from the joists, creating gaps. Where gaps are severe or appear above several neighboring joists, the framing may need reinforcement, but isolated gapping can usually be remedied with hardwood shims. Apply a small amount of wood glue to the shim and squirt some glue into the gap. Using a hammer, tap the shim into place until it's snug. Shimming too much will widen the gap, so be careful. Allow the glue to dry before walking on the floor.

Eliminating Stair Squeaks

This staircase has center stringers to help support the treads. The 2 × 4s nailed between the outside stringers and the wall studs serve as spacers that allow room for the installation of skirt boards and wall finishes.

Like floors, stairs squeak when the lumber becomes warped or loose boards rub together. The continual pounding of foot traffic takes its toll on even the best built staircases. An unstable staircase is as unsafe as it is unattractive. Problems related to the structure of a staircase, such as severe sagging, twisting, or slanting, should be left to a professional. However, you can easily complete many common repairs.

Squeaks are usually caused by movement between the treads and risers, which can be alleviated from above or below the staircase.

Tools & Materials ▸

- Drill
- Screwdriver
- Hammer
- Utility knife
- Nail set
- Wood screws
- Wood putty
- Caulk gun
- Hardwood shims
- Wood plugs
- Wood glue
- Quarter-round molding
- Finish nails
- Construction adhesive

How to Eliminate Squeaks from Below the Stairs

Glue wood blocks to the joints between the treads and risers with construction adhesive. Once the blocks are in place, drill pilot holes and fasten them to the treads and risers with wood screws. If the risers overlap the back edges of the treads, drive screws through the risers and into the treads to bind them together.

Fill the gaps between stair parts with tapered hardwood shims. Coat the shims with wood glue and tap them into the joints between treads and risers until they're snug. Shimming too much will widen the gap. Allow the glue to dry before walking on the stairs.

How to Eliminate Squeaks from Above the Stairs

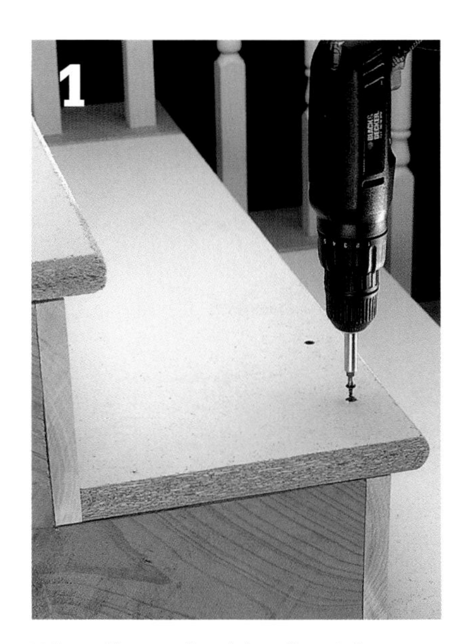

When the underside of a staircase is inaccessible, silence noisy stairs from above. Drill pilot holes and drive screws down through stair treads into the risers. Countersink the screws and fill the holes with putty or wood plugs.

Support the joints between treads and risers by attaching quarter-round molding. Drill pilot holes and use finish nails to fasten the molding. Set the nails with a nail set.

Tap glued wood shims under loose treads to keep them from flexing. Use a block to prevent splitting, and drive the shim just until it's snug. When the glue dries, cut the shims flush, using a utility knife.

Replacing a Broken Stair Tread

A broken stair tread is hazardous because we often don't look at steps as we climb them. Replace a broken step right away. The difficulty of this job depends on the construction of your staircase and the accessibility of the underside. It's better to replace a damaged tread than to repair it. A patch could create an irregular step that surprises someone unfamiliar with it.

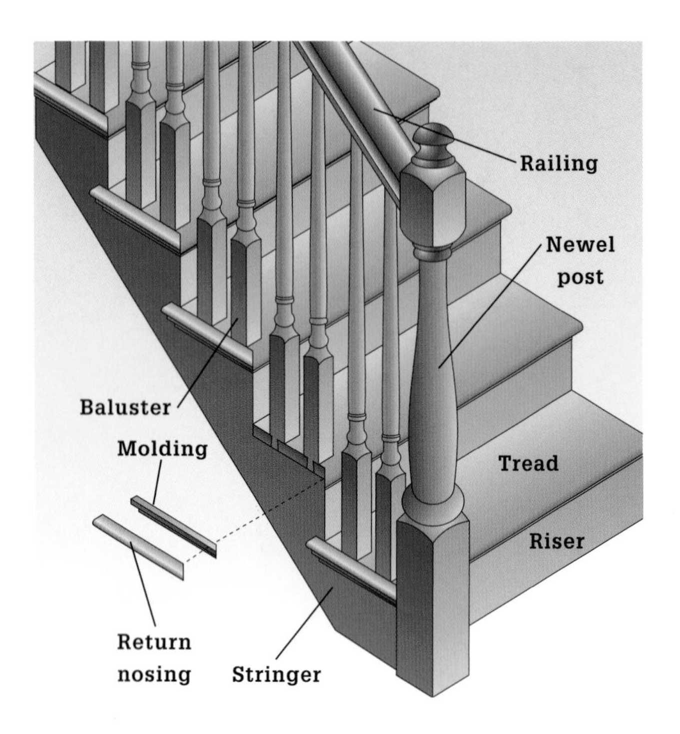

Tools & Materials ▸

- Flat pry bar
- Hammer
- Combination square
- Circular saw
- Drill
- Nail set
- Caulk gun
- Stair tread
- Construction adhesive
- Screws
- Wood putty
- Finish nails
- Nail set

How to Replace a Broken Stair Tread

Carefully remove any decorative elements attached to the tread. Pull up carpeting and roll it aside. Remove trim pieces on or around the edges of the tread. Remove the balusters by detaching the top ends from the railing and separating the joints in the tread. Some staircases have a decorative hardwood cap inlaid into each tread. Remove these with a flat pry bar, taking care to pry from underneath the cap to avoid marring the exposed edges.

If possible, hammer upward from underneath the stairs to separate the tread from the risers and stringers. Otherwise, use a hammer and pry bar to work the tread loose, pulling nails as you go. Once the tread is removed, scrape the exposed edges of the stringers to remove old glue and wood fragments.

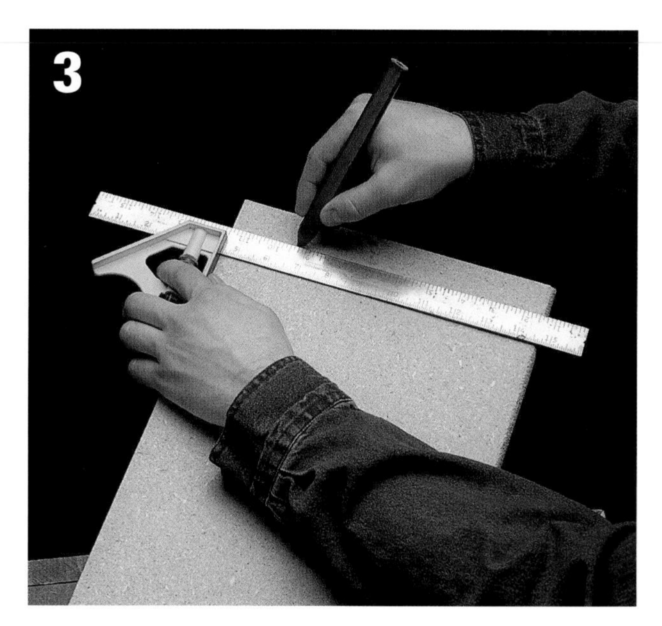

Measure the length for the new tread and mark it with a combination square so the cut end will be square and straight. If the tread has a milled end for an inlay, cut from the plain end. Cut the new tread to size, using a circular saw, and test-fit it carefully.

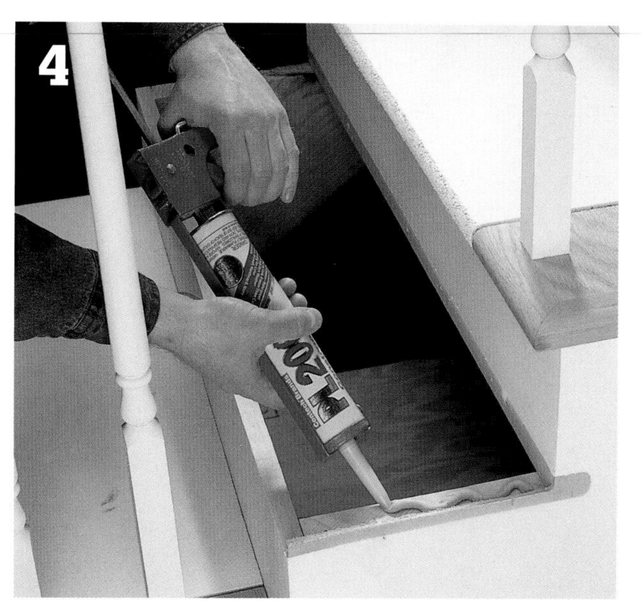

Apply a bead of construction adhesive to the exposed tops of the stringers. The adhesive will strengthen the bond between the tread and stringer and will cushion the joint, preventing the parts from squeaking.

Set the tread in place. If you have access to the step from underneath, secure the tread to the riser above it by driving screws through the riser into the back edge of the tread. To fasten it from the top side, drill and countersink pilot holes and drive two or three screws through the tread into the top edge of each stringer. Also drive a few screws along the front edge of the tread into the riser below it. Fill the screw holes in the tread with wood putty or plugs.

Reinstall any decorative elements, using finish nails. Set the nails with a nail set. Reinstall the balusters, if necessary.

Replacing Trim Moldings

There's no reason to let damaged trim moldings detract from the appearance of a well-maintained room. With the right tools and a little attention to detail, you can replace or repair them quickly and easily.

Home centers and lumber yards sell many styles of moldings, but they may not stock moldings found in older homes. If you have trouble finding duplicates, check salvage yards in your area. They sometimes carry styles no longer manufactured. You can also try combining several different moldings to duplicate a more elaborate version.

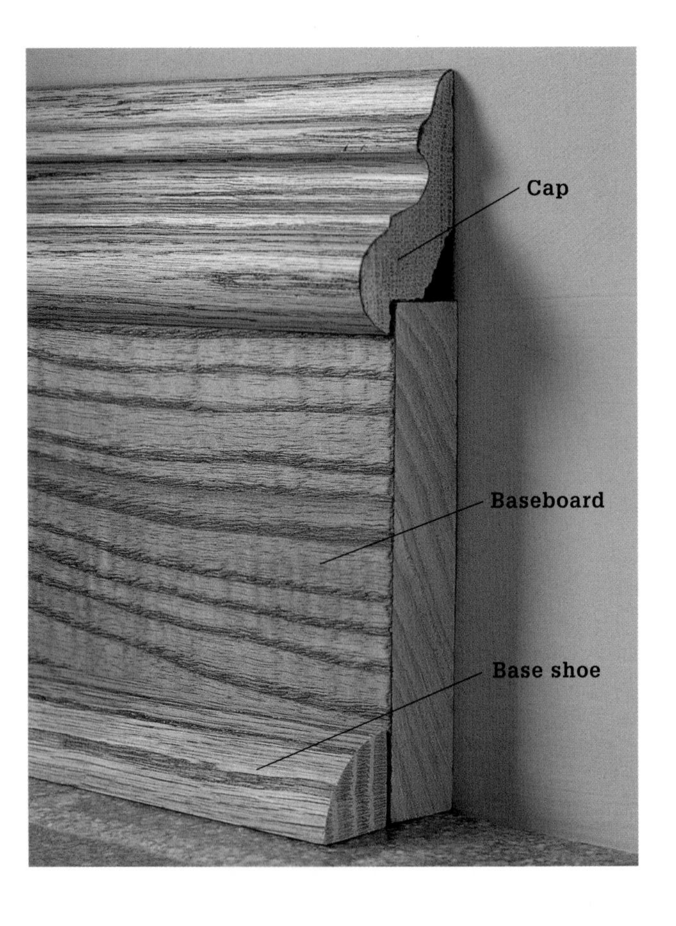

Tools & Materials ▸

- Flat pry bars (2)
- Coping saw
- Miter saw
- Drill
- Hammer
- Nail set
- Wood scraps
- Replacement moldings
- 2d, 4d, and 6d finish nails
- Wood putty

How to Remove Damaged Trim

Even the lightest pressure from a pry bar can damage wallboard or plaster, so use a large, flat scrap of wood to protect the wall. Insert one bar beneath the trim and work the other bar between the baseboard and the wall. Force the pry bars in opposite directions to remove the baseboard.

To remove baseboards without damaging the wall, use leverage rather than force. Pry off the base shoe first, using a flat pry bar. When you feel a few nails pop, move farther along the molding and pry again.

How to Install Baseboards

Start at an inside corner by butting one piece of baseboard securely into the corner. Drill pilot holes, then fasten the baseboard with two 6d finish nails, aligned vertically, at each wall stud. Cut a scrap of baseboard so the ends are perfectly square. Cut the end of the workpiece square. Position the scrap on the back of the workpiece so its back face is flush with the end of the workpiece. Trace the outline of the scrap onto the back of the workpiece.

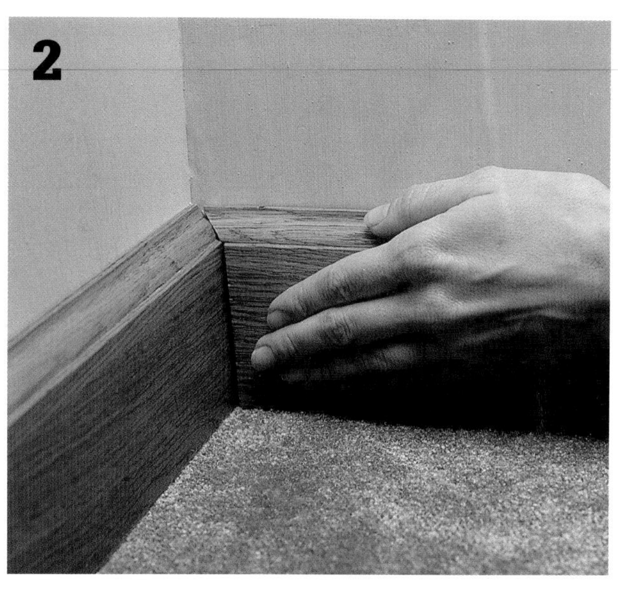

Cut along the outline on the workpiece with a coping saw, keeping the saw perpendicular to the baseboard face. Test-fit the coped end. Recut it, if necessary.

To cut the baseboard to fit at outside corners, mark the end where it meets the outside wall corner. Cut the end at a 45° angle, using a power miter saw. Lock-nail all miter joints by drilling a pilot hole and driving 4d finish nails through each corner.

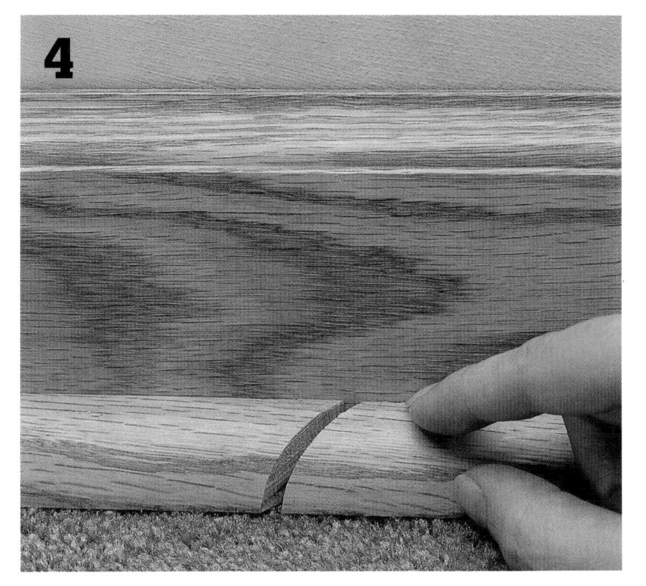

Install base shoe molding along the bottom of the baseboards. Make miter joints at inside and outside corners, and fasten base shoe with 2d finish nails. Whenever possible, complete a run of molding using one piece. For long spans, join molding pieces by mitering the ends at parallel 45° angles. Set nail heads below the surface using a nail set, and then fill the holes with wood putty.

Repairing Hardwood

A darkened, dingy hardwood floor may only need a thorough cleaning to reveal an attractive, healthy finish. If you have a fairly new or prefinished hardwood floor, check with the manufacturer or flooring installer before applying any cleaning products or wax. Most prefinished hardwood, for example, should not be waxed.

Water and other liquids can penetrate deep into the grain of hardwood floors, leaving dark stains that are sometimes impossible to remove by sanding. Instead, try bleaching the wood with oxalic acid, available in crystal form at home centers or paint stores. When gouges, scratches, and dents aren't bad enough to warrant replacing a floorboard, repair the damaged area with a latex wood patch that matches the color of your floor.

Identify surface finishes using solvents. In an inconspicuous area, rub in different solvents to see if the finish dissolves, softens, or is removed. Denatured alcohol removes shellac, while lacquer thinner removes lacquer. If neither of those work, try nail polish remover containing acetone, which removes varnish but not polyurethane.

Tools & Materials ▸

- Vacuum
- Buffing machine
- Hammer
- Nail set
- Putty knife
- Cloths
- Hardwood cleaning kit
- Paste wax
- Rubber gloves
- Oxalic acid
- Vinegar
- Wood restorer
- Latex wood patch
- Sandpaper
- Drill
- Spade bit
- Circular saw
- Chisel
- Hammer
- Caulk gun
- Spine-shank flooring nails
- Nail set
- Wood putty

How to Clean & Renew Hardwood

Vacuum the entire floor. Mix hot water and dishwashing detergent that doesn't contain lye, trisodium phosphate, or ammonia. Working on 3-ft.-square sections, scrub the floor with a brush or nylon scrubbing pad. Wipe up the water and wax with a towel before moving to the next section.

If the water and detergent don't remove the old wax, use a hardwood floor cleaning kit. Use only solvent-type cleaners, as some water-based products can blacken wood. Apply the cleaner following the manufacturer's instructions.

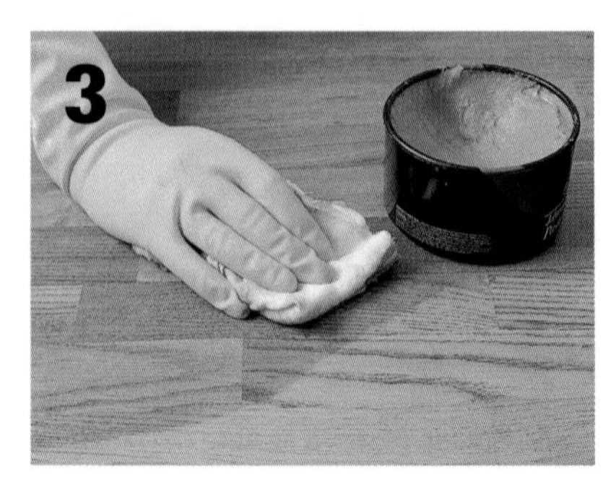

When the floor is clean and dry, apply a high-quality floor wax. Paste wax is more difficult to apply than liquid floor wax, but it lasts much longer. Apply the wax by hand, then polish the floor with a rented buffing machine fitted with synthetic buffing pads.

How to Remove Stains

Remove the finish by sanding the stained area with sandpaper. In a disposable cup, dissolve the recommended amount of oxalic acid crystals in water. Wearing rubber gloves, pour the mixture over the stained area, taking care to cover only the darkened wood.

Let the liquid stand for one hour. Repeat the application, if necessary. Wash with 2 tablespoons borax dissolved in one pint water to neutralize the acid. Rinse with water, and let the wood dry. Sand the area smooth.

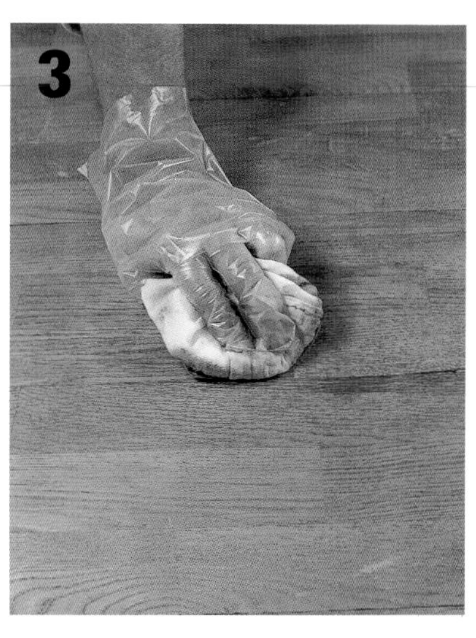

Apply several coats of wood restorer until the bleached area matches the finish of the surrounding floor.

How to Patch Scratches & Small Holes

Before filling nail holes, make sure the nails are securely set in the wood. Use a hammer and nail set to drive loose nails below the surface. Apply wood patch to the damaged area, using a putty knife. Force the compound into the hole by pressing the knife blade downward until it lies flat on the floor.

Scrape excess compound from the edges, and allow the patch to dry completely. Sand the patch flush with the surrounding surface. Using fine-grit sandpaper, sand in the direction of the wood grain.

Apply wood restorer to the sanded area until it blends with the rest of the floor.

Replacing a Section of a Damaged Floorboard

When solid hardwood floorboards are beyond repair, they need to be carefully cut out and replaced with boards of the same width and thickness. Replace whole boards whenever possible. If a board is long, or if part of its length is inaccessible, draw a cutting line across the face of the board, and tape behind the line to protect the section that will remain.

How to Replace Damaged Floorboards

Draw a rectangle around the damaged area. Determine the minimal number of boards to be removed. To avoid nails, be sure to draw the line ¾" inside the outermost edge of any joints.

Determine the depth of the boards to be cut. With a drill and ¾"-wide spade bit, slowly drill through a damaged board. Drill until you see the top of the subfloor. Measure the depth. A common depth is ⅝" or ¾". Set your circular saw to this depth.

To prevent boards from chipping, place masking tape or painter's tape along the outside of the pencil lines. To create a wood cutting guide, tack a straight wood strip inside the damaged area (for easy removal, allow nails to slightly stick up). Set back the guide the distance between the saw blade and the guide edge of the circular saw.

(continued)

Align the circular saw with the wood cutting guide. Turn on the saw. Lower the blade into the cutline. Do not cut the last ¼" of the corners. Remove cutting guide. Repeat with other sides.

Complete the cuts. Use a hammer and sharp chisel to completely loosen the boards from the subfloor. Make sure the chisel's beveled side is facing the damaged area for a clean edge.

Remove split boards. Use a scrap 2 × 4 block for leverage and to protect the floor. With a hammer, tap a pry bar into and under the split board. Most boards pop out easily, but some may require a little pressure. Remove exposed nails with the hammer claw.

Use a chisel to remove the 2 remaining strips. Again, make sure the bevel side of the chisel is facing the interior of the damaged area. Set any exposed nails with your nail set.

Cut new boards. Measure the length and width of the area to be replaced. Place the new board on a sawhorse, with the section to be used hanging off the edge. Draw a pencil cutline. With saw blade on waste side of mark, firmly press the saw guide against the edge of a framing square. Measure each board separately.

Use a mallet or hammer to gently tap the groove of the new board into the tongue of the existing board. To protect the tongue of the new board, use a scrap 2 × 4 or a manufacturer's block as a hammering block.

(continued)

Pick a drill bit with a slightly smaller diameter than an 8-penny finish nail, and drill holes at a 45° angle through the corner of the replacement piece's tongue every 3" to 4" along the new board. Hammer a 1½"-long, 8-penny finish nail through the hole into the subfloor. Use a nail set to countersink nails. Repeat until the last board.

Lay the last board face down onto a protective 2 × 4 and use a sharp chisel to split off the lower lip. This allows it to fit into place.

To install the last board, hook the tongue into the groove of the old floor and then use a soft mallet to tap the groove side down into the previous board installed.

Drill pilot holes angled outward: two side-by-side holes about ½" from the edges of each board, and one hole every 12" along the groove side of each board. Drive 1½"-long, 8-penny finish nails through the holes. Set nails with a nail set. Fill holes with wood putty.

Once the putty is dry, sand the patch smooth with fine-grit sandpaper. Feather-sand neighboring boards. Vacuum and wipe the area with a clean cloth. Apply matching wood stain or restorer; then apply 2 coats of matching finish.

Variation: Staggered Pattern ▸

To make a more seamless patch, install a staggered pattern. Follow the basic instructions for the rectangular patch, with the following exceptions:

Drill a series of holes into the damaged boards with a ⅝" Forstner bit or hole saw bit. Mark the drilling depth with masking tape so you do not drill into the subfloor.

Split and pry out boards. Set the circular saw depth to floorboard thickness. Split damaged boards lengthwise. Cut outward from the center until the cuts intersect the holes. Chisel out wood between holes. Using a 2 × 4 block for leverage, pry out boards.

Install new boards. Place outermost boards first, working your way in. Use a 2 × 4 scrap hammering block and hammer to tap the new board into place. Lay the last board on a protective 2 × 4 and use a sharp chisel to split off the lower lip. See Step 12 for installing last board.

Replacing Small Sections of Wood Floors

Splinters often appear in floors that are dried out and brittle. When hardwood floors are damaged by high heels or pushed chair legs, a portion of the grain may dislodge; because the grain of wood runs only in one direction, it splinters rather than simply creating a hole. Floorboards that have splinters or gouges don't necessarily have to be replaced; a splinter can be reattached with some glue, and a hole can be filled with some wood putty.

How to Repair Splinters

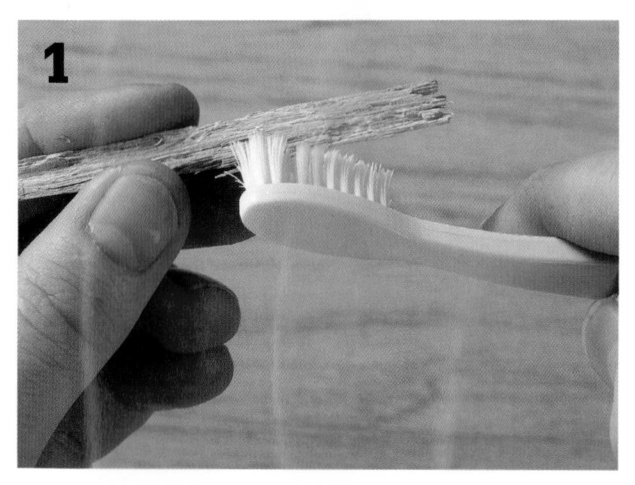

If you still have the splintered piece of wood, but it has been entirely dislodged from the floor, it's a good bet that the hollowed space left by the splinter has collected a lot of dirt and grime. Combine a 1:3 mixture of distilled white vinegar and water in a bucket. Dip an old toothbrush into the solution and use it to clean out the hole left in the floor. While you're at it, wipe down the splinter with the solution, too. Allow the floor and splinter to thoroughly dry.

If the splinter is large, apply wood glue to the hole and splinter. Use a Q-Tip or toothpick to apply small amounts of wood glue under smaller splinters. Soak the Q-Tip in glue; you don't want Q-Tip fuzz sticking out of your floor once the glue dries.

Press the splinter back into place. To clean up the excess glue, use a slightly damp, lint-free cloth. Do not oversoak the cloth with water.

Allow the adhesive to dry. Cover the patch with wax paper and a couple of books. Let the adhesive dry overnight.

How to Patch Small Holes in Wood Floors

Repair small holes with wood putty. Use putty that matches the floor color. Force the compound into the hole with a putty knife. Continue to press the putty in this fashion until the depression in the floor is filled. Scrape excess compound from the area. Use a damp, lint-free cloth while the putty is still wet to smooth the top level with the surrounding floor. Allow to dry.

Sand the area with fine (100- to 120-grit) sandpaper. Sand with the wood grain so the splintered area is flush with the surrounding surface. To better hide the repair, feather sand the area. Wipe up dust with a slightly damp cloth.

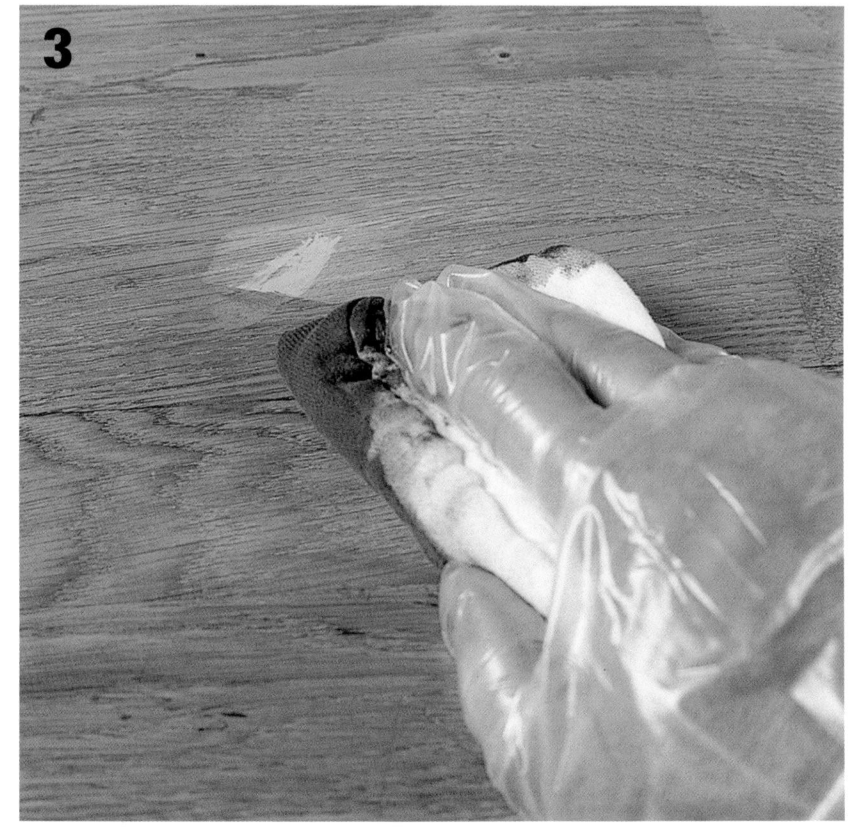

With a clean, lint-free cloth, apply a matching stain (wood sealer or "restorer") to the sanded area. Read the label on the product to make sure it is appropriate for sealing wood floors. Work in the stain until the patched area blends with the rest of the floor. Allow area to completely dry. Apply two coats of finish. Be sure the finish is the same as that which was used on the surrounding floor.

Replacing Large Sections of Wood Floors

When an interior wall or section of wall has been removed during remodeling, you'll need to patch gaps in the flooring where the wall was located. There are several options for patching floors, depending on your budget and the level of your do-it-yourself skills.

If the existing flooring shows signs of wear, consider replacing the entire flooring surface. Although it can be expensive, an entirely new floor covering will completely hide any gaps in the floor and provide an elegant finishing touch for your remodeling project.

If you choose to patch the existing flooring, be aware that it's difficult to hide patched areas completely, especially if the flooring uses unique patterns or finishes. A creative solution is to intentionally patch the floor with material that contrasts with the surrounding flooring (opposite page).

How to Replace Large Sections of Wood Floors

When patching a wood-strip floor, one option is to remove all of the floor boards that butt against the flooring gap using a pry bar and replace them with boards cut to fit. This may require you to trim the tongues from some tongue-and groove floorboards. Sand and refinish the entire floor so the new boards match the old.

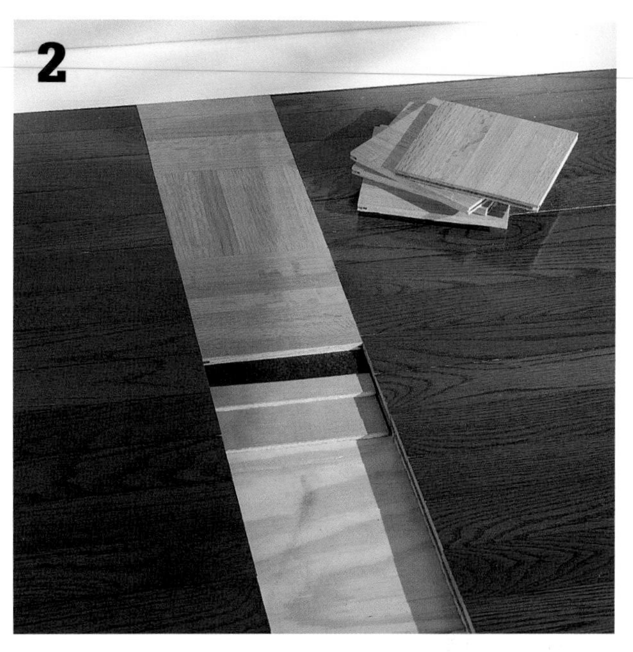

Build up the subfloor in the patch area, using layers of thin plywood and building paper, so the new surface will be flush with the surrounding flooring. You may need to experiment with different combinations of plywood and paper to find the right thickness.

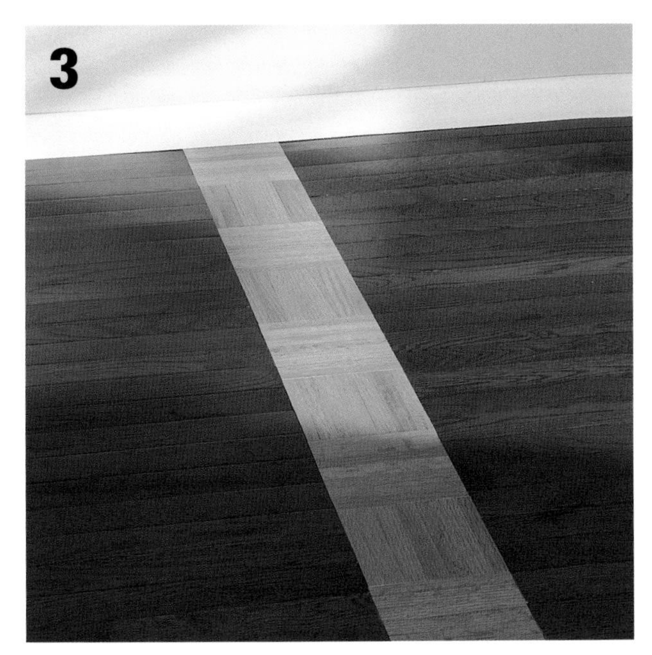

Fill gaps in floors with materials that have a contrasting color and pattern. For wood floors, parquet tiles are an easy and inexpensive choice. You may need to widen the gap with a circular saw set to the depth of the wood covering to make room for the contrasting tiles.

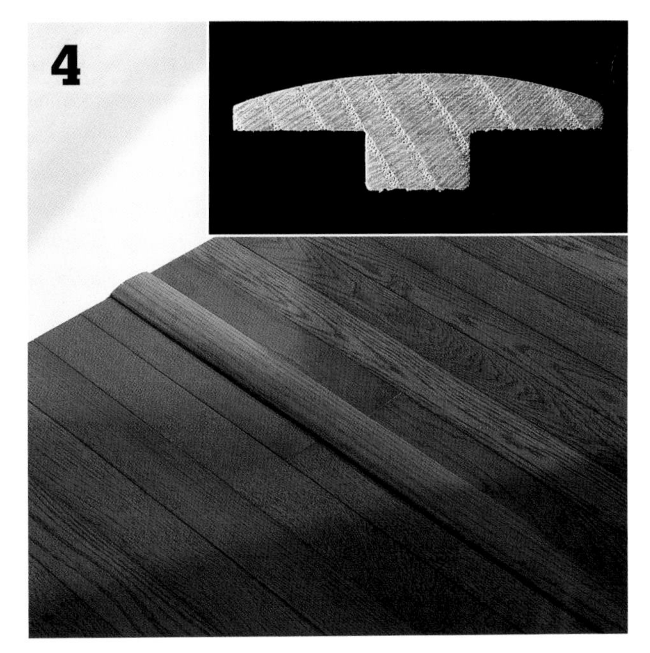

A quick, inexpensive solution is to install T-molding to bridge a gap in a wood strip floor. T-moldings are especially useful when the surrounding boards run parallel to the gap. T-moldings are available in several widths and can be stained to match the flooring.

Repairing Vinyl

Repair methods for vinyl flooring depend on the type of floor as well as the type of damage. With sheet vinyl, you can fuse the surface or patch in new material. With vinyl tile, it's best to replace the damaged tiles.

Small cuts and scratches can be fused permanently and nearly invisibly with liquid seam sealer, a clear compound that's available wherever vinyl flooring is sold. For tears or burns, the damaged area can be patched. If necessary, remove vinyl from a hidden area, such an the inside of a closet or under an appliance, to use as patch material.

When vinyl flooring is badly worn or the damage is widespread, the only answer is complete replacement. Although it's possible to add layers of flooring in some situations, evaluate the options carefully. Be aware that the backing of older vinyl tiles made of asphalt may contain asbestos fibers. Consult a professional for their removal.

Tools & Materials ▸

- Carpenter's square
- Utility knife
- Putty knife
- Heat gun
- J-roller
- Notched trowel
- Marker
- Masking tape
- Scrap of matching flooring
- Mineral spirits
- Floor covering adhesive
- Wax paper
- Liquid seam sealer

How to Patch Sheet Vinyl

Measure the width and length of the damaged area. Place the new flooring remnant on a surface you don't mind making some cuts on—like a scrap of plywood. Use a carpenter's square for cutting guidance. Make sure your cutting size is a bit larger than the damaged area.

Lay the patch over the damaged area, matching pattern lines. Secure the patch with duct tape. Using a carpenter's square as a cutting guide, cut through the new vinyl (on top) and the old vinyl (on bottom). Press firmly with the knife to cut both layers.

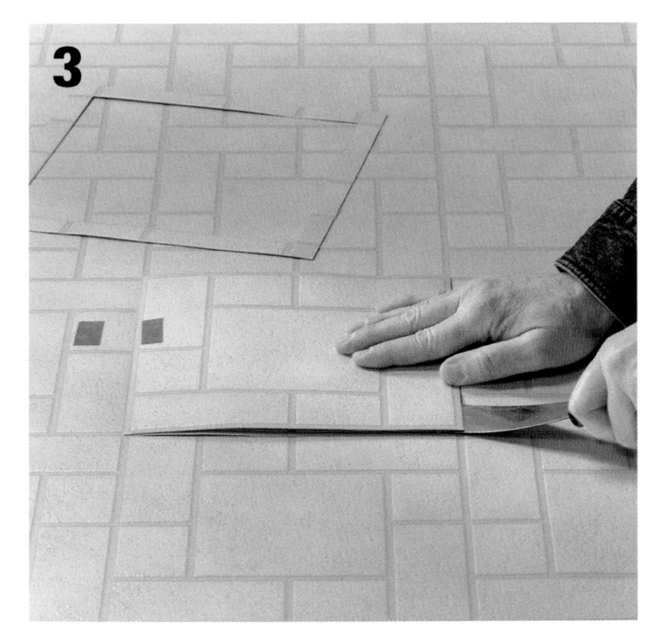

Use tape to mark one edge of the new patch with the corresponding edge of the old flooring as placement marks. Remove the tape around the perimeter of the patch and lift up.

Soften the underlying adhesive with an electric heat gun and remove the damaged section of floor. Work from edges in. When the tile is loosened, insert a putty knife and pry up the damaged area.

(continued)

Scrape off the remaining adhesive with a putty knife or chisel. Work from the edges to the center. Dab mineral spirits (or Goo Gone) or spritz warm water on the floor to dissolve leftover goop, taking care not to use too much; you don't want to loosen the surrounding flooring. Use a razor-edged scraper (flooring scraper) to scrape to the bare wood underlayment.

Apply adhesive to the patch, using a notched trowel (with ⅛" V-shaped notches) held at a 45° angle to the back of the new vinyl patch.

Set one edge of the patch in place. Lower the patch onto the underlayment. Press into place. Apply pressure with a J-roller or rolling pin to create a solid bond. Start at the center and work toward the edges, working out air bubbles. Wipe up adhesive that oozes out the sides with a clean, damp cloth or sponge.

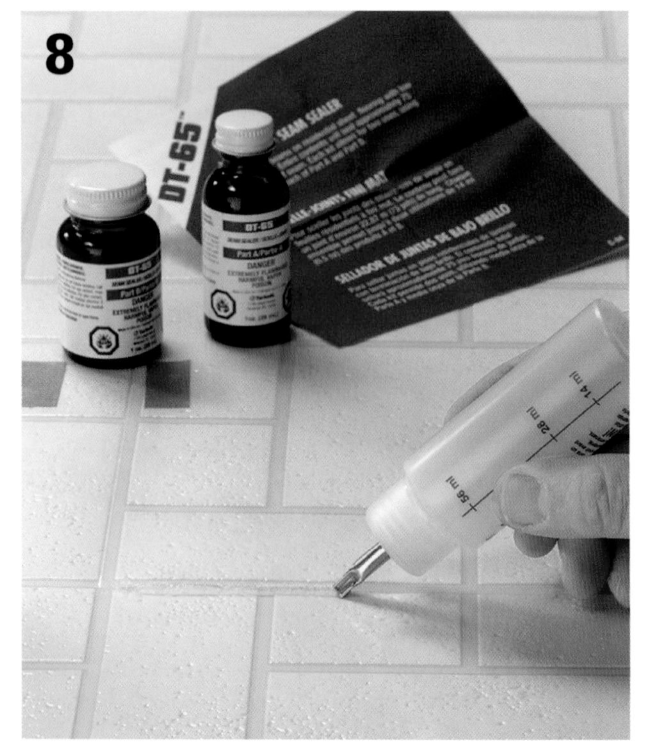

Let the adhesive dry overnight. Use a soft cloth dipped in lacquer thinner to clean the area. Mix the seam sealer according to the manufacturer's directions. Use an applicator bottle to apply a thin bead of sealer onto the cutlines.

How to Replace Resilient Tile

Use an electric heat gun to warm the damaged tile and soften the underlying adhesive. Keep the heat source moving so you don't melt the tile. When an edge of the tile begins to curl, insert a putty knife to pry up the loose edge until you can remove the tile. *Note: If you can clearly see the seam between tiles, first score around the tile with a utility knife. This prevents other tiles from lifting.*

Scrape away remaining adhesive with a putty knife or, for stubborn spots, a floor scraper. Work from the edges to the center so that you don't accidentally scrape up the adjacent tiles. Use mineral spirits to dissolve leftover goop. Take care not to allow the mineral spirits to soak into the floor under adjacent tiles. Vacuum up dust, dirt, and adhesive. Wipe clean.

When the floor is dry, use a notched trowel—with $\frac{1}{8}$" V-shaped notches—held at a 45° angle to apply a thin, even layer of vinyl tile adhesive onto the underlayment. *Note: Only follow this step if you have dry-back tiles.*

Set one edge of the tile in place. Lower the tile onto the underlayment and then press it into place. Apply pressure with a J-roller to create a solid bond, starting at the center and working toward the edge to work out air bubbles. If adhesive oozes out the sides, wipe it up with a damp cloth or sponge. Cover the tile with wax paper and some books, and let the adhesive dry for 24 hours.

Repairing Carpet

Burns and stains are the most common carpeting problems. You can clip away the burned fibers of superficial burns using small scissors. Deeper burns and indelible stains require patching by cutting away and replacing the damaged area.

Another common problem, addressed on the opposite page, is carpet seams or edges that have come loose. You can rent the tools necessary for fixing this problem.

Tools & Materials ▸

- Cookie-cutter tool
- Knee kicker
- 4" wallboard knife
- Utility knife
- Seam iron
- Replacement carpeting
- Double-face carpet tape
- Seam adhesive
- Heat-activated seam tape
- Boards
- Weights

How to Repair Spot Damage

Remove extensive damage or stains with a "cookie-cutter" tool, available at carpeting stores. Press the cutter down over the damaged area and twist it to cut away the carpet.

Using the cookie-cutter tool again, cut a replacement patch from scrap carpeting. Insert double-face carpet tape under the cutout, positioning the tape so it overlaps the patch seams.

Press the patch into place. Make sure the direction of the nap or pattern matches the existing carpet. To seal the seam and prevent unraveling, apply seam adhesive to the edges of the patch.

How to Restretch Loose Carpet

Adjust the knob on the head of the knee kicker so the prongs grab the carpet backing without penetrating through the padding. Starting from a corner or near a point where the carpet is firmly attached, press the knee kicker head into the carpet, about 2" from the wall.

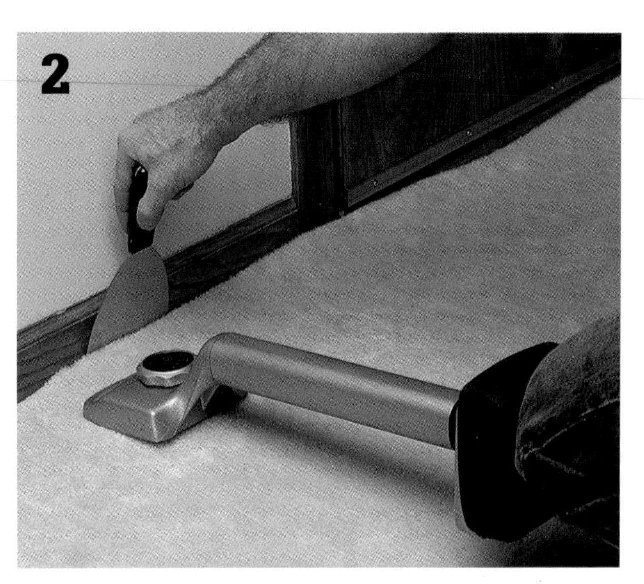

Thrust your knee into the cushion of the knee kicker to force the carpet toward the wall. Tuck the carpet edge into the space between the wood strip and the baseboard using a 4" wallboard knife. If the carpet is still loose, trim the edge with a utility knife and stretch it again.

How to Glue Loose Seams

Remove the old tape from under the carpet seam. Cut a strip of new seam tape and place it under the carpet so it's centered along the seam with the adhesive facing up. Plug in the seam iron and let it heat up.

Pull up both edges of the carpet and set the hot iron squarely onto the tape. Wait about 30 seconds for the glue to melt. Move the iron about 12" farther along the seam. Quickly press the edges of the carpet together into the melted glue behind the iron. Separate the pile to make sure no fibers are stuck in the glue and the seam is tight. Place weighted boards over the seam to keep it flat while the glue sets. Remember, you have only 30 seconds to repeat the process.

Repairing Ceramic Tile

Although ceramic tile is one of the hardest floor coverings, tiles sometimes become damaged and need to be replaced. Major cracks in grout joints indicate that floor movement has caused the adhesive layer beneath the tile to deteriorate. The adhesive layer must be replaced in order to create a permanent repair.

Any time you remove tile, check the underlayment. If it's no longer smooth, solid, and level, repair or replace it before replacing the tile. When removing grout or damaged tiles, be careful not to damage surrounding tiles. Always wear eye protection when working with a hammer and chisel.

Tools & Materials ▸

- Hammer
- Cold chisel
- Eye protection
- Putty knife
- Square-notched trowel
- Rubber mallet
- Level
- Needlenose pliers
- Screwdriver
- Grout float mix
- Thin-set mortar
- Replacement tile
- Tile spacers
- Grout
- Bucket
- Grout pigment
- Grout sealer
- Grout sponge
- Floor-leveling compound

How to Replace Ceramic Tiles

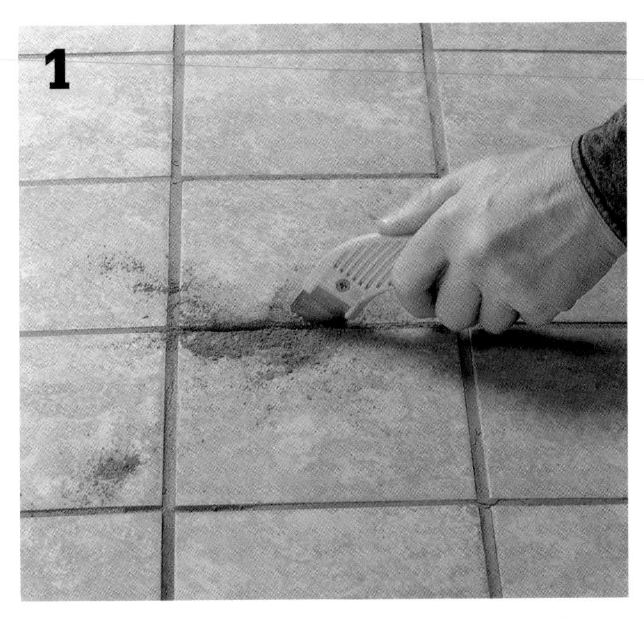

With a carbide-tipped grout saw, apply firm but gentle pressure across the grout until you expose the unglazed edges of the tile. Do not scratch the glazed tile surface. If the grout is stubborn, use a hammer and screwdriver to first tap the tile (Step 2).

If the tile is not already cracked, use a hammer to puncture the tile by tapping a nail set or center punch into it. Alternatively, if the tile is significantly cracked, use a chisel to pry up the tile.

Insert a chisel into one of the cracks and gently tap the tile. Start at the center and chip outward so you don't damage the adjacent tiles. Be aware that cement board looks a lot like mortar when you're chiseling. Remove and discard the broken pieces.

Use a putty knife to scrape away old thinset adhesive; use a chisel for poured mortar installation. If the underlayment is covered with metal lath, you won't be able to get the area smooth; just clean it out the best you can. Once the adhesive is scraped from the underlayment, smooth the rough areas with sandpaper. If there are gouges in the underlayment, fill them with epoxy-based thinset mortar (for cementboard) or a floor-leveling compound (for plywood). Allow the area to dry completely.

(continued)

Use a ¼" notched trowel to apply thinset adhesive to the back of the replacement tile. Set the tile down into the space, and use plastic spacers around the tile to make sure it is centered within the opening.

Set the tile in position and press down until it is even with the adjacent tiles. Twist it a bit to get it to sit down in the mortar. Use a mallet or hammer and a block of wood covered with cloth or a carpet scrap (a "beater block") to lightly tap on the tile, setting it into the adhesive. Use a level or other straight surface to make sure the tile is level with the surrounding tiles. If necessary, continue to tap the tile until it's flush with the rest of the surrounding tiles.

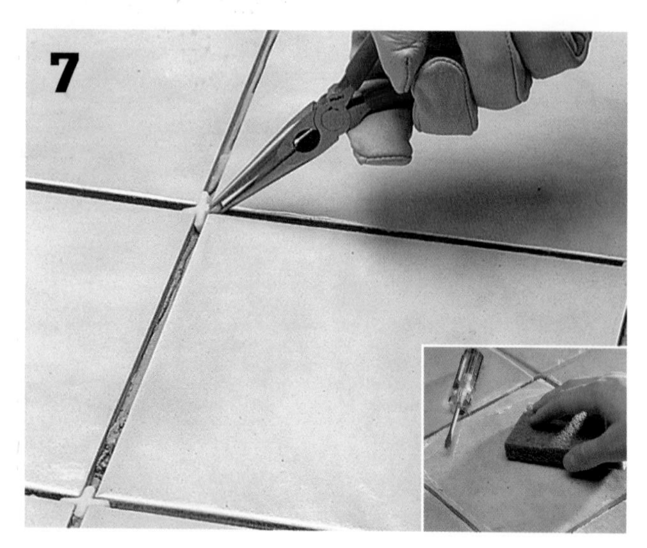

Remove the spacers with needlenose pliers. Get the mortar or thinset adhesive out of the grout joints with a small screwdriver and a cloth. Also, wipe away any adhesive from the surface of the tiles, using a wet sponge. Let the area dry for 24 hours (or according to the manufacturer's recommendations).

Use a putty knife to apply grout to the joints. Fill in low spots by applying and smoothing extra grout with your finger. Use the round edge of a toothbrush handle to create a concave grout line, if desired. You must now grout the joint.

Repairing Concrete

When patching concrete it's important to clean first. Concrete is formed by a chemical reaction between Portland cement and water, which is called "hydration." The interaction creates tiny crystals that interlock with one another, binding the sand and gravel aggregates into its structure. If there's dirt in the repair, the crystals bond to the dirt instead of the old concrete and the repair flakes out over time. So long as the cleaning steps in this project are taken seriously, the repairs are strong and long-lasting.

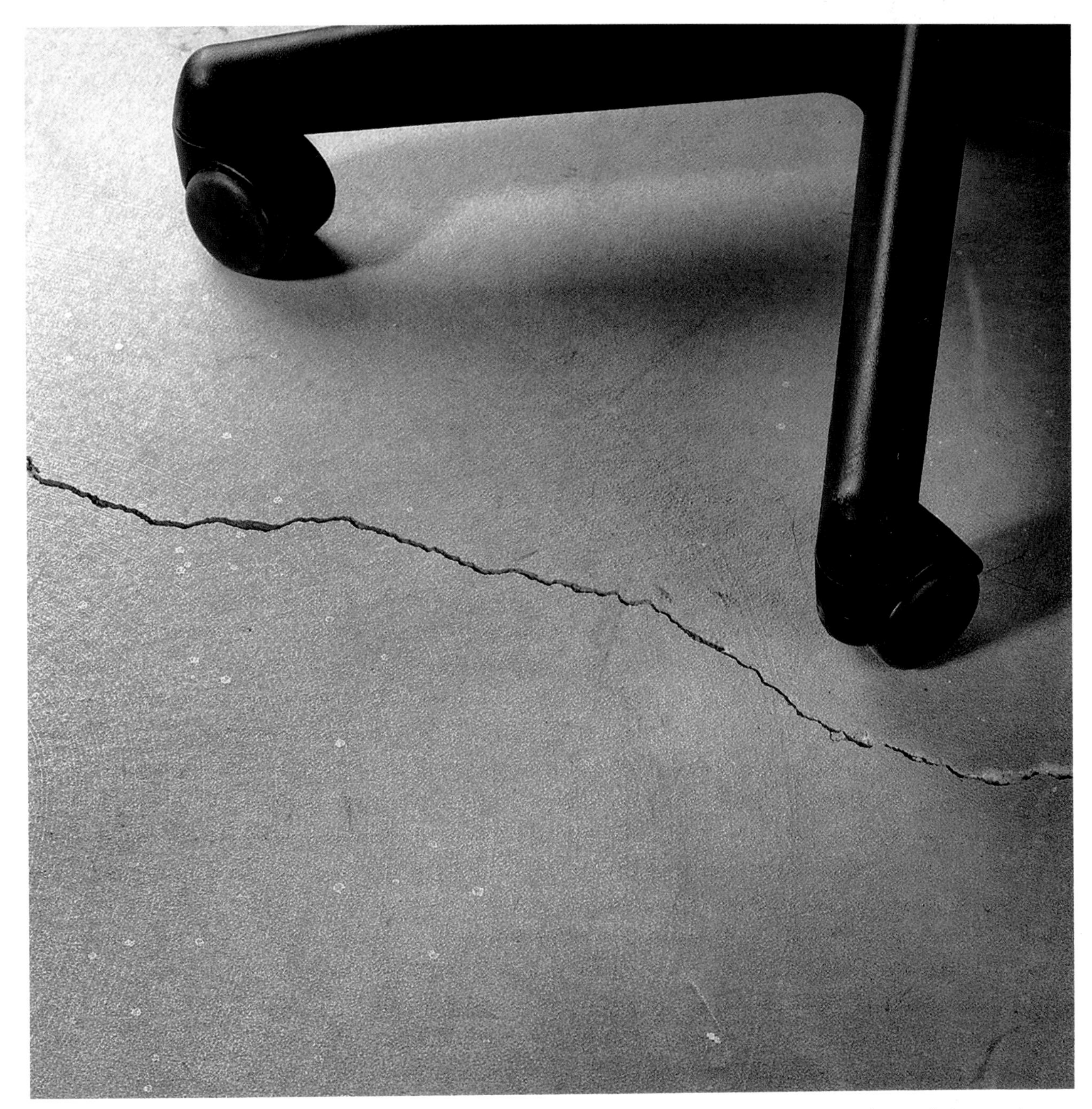

Although concrete is very durable, it can still be chipped by heavy objects or cracked due to abrupt changes in temperature or house shifting.

How to Repair Cracks in Concrete

Use a concrete chisel (called a "cold chisel") and a heavy hammer or mallet to deepen the edges of the damaged area until the outer edges are at least ⅛" thick. Most cracks and depressions in concrete floors are deeper in the center and are tapered at the edges; the feather thin material around the perimeter of the hole is liable to peel or flake off, which results in an unstable surface for a patch.

Clean out the area to be patched, using a wire brush or portable drill with a wire wheel attachment. Be sure to remove all dirt and loose material from the area to be patched. This step will also roughen the edges a bit, creating a better bond.

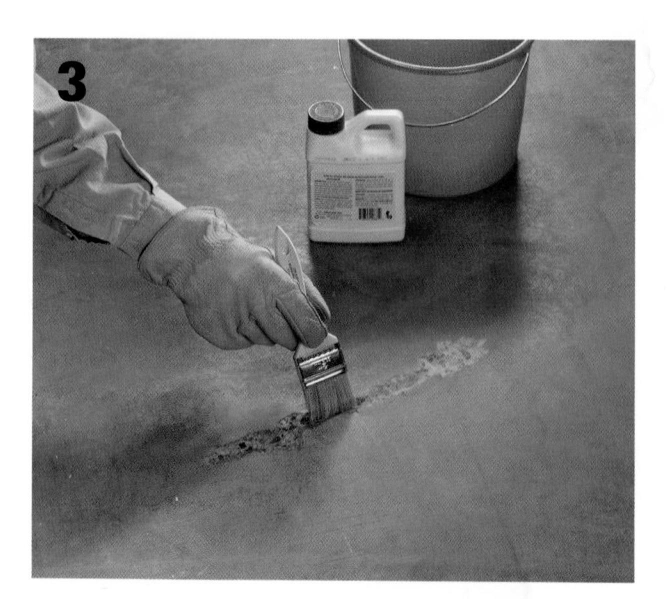

A bonding agent (also called a bonding "adhesive") helps to chemically bond the patch material to the existing concrete, making the repair material less likely to loosen or dislodge. Apply a thin layer of bonding adhesive to the entire repair area with a paintbrush. Some bonding agents need to be applied to a wet surface, others should not. Follow the directions carefully.

Mix your concrete patching compound with clean water until all of the material is thoroughly wet and all of the lumps are worked out. Most mixing compounds start to set within 10–20 minutes. (Inset) Use a trowel to compact the material into the area being repaired until it is slightly raised above the surface of the surrounding concrete. If the hole is deeper than ¼", allow each layer to dry before applying the next layer. This prevents the top from drying out and shrinking while the lower areas are wet, which could cause re-cracking.

Use the edge of the trowel to smooth the surface, removing any excess material. Slide the trowel back and forth on its edge, while also pulling the excess material toward you, until it is past the edge of the area you're working on. Scoop it up with the trowel and discard.

Finishing work. Slightly raise the flat face of the steel finishing trowel and smooth the patching material until it is even with the adjoining surfaces, creating a seamless repair. Keep the trowel clean and damp to prevent the mix from gumming up the trowel. Finishing is an art and takes practice, so keep trying. Remember that the patching compound will be easier to work with when it's at a slightly stiff consistency.

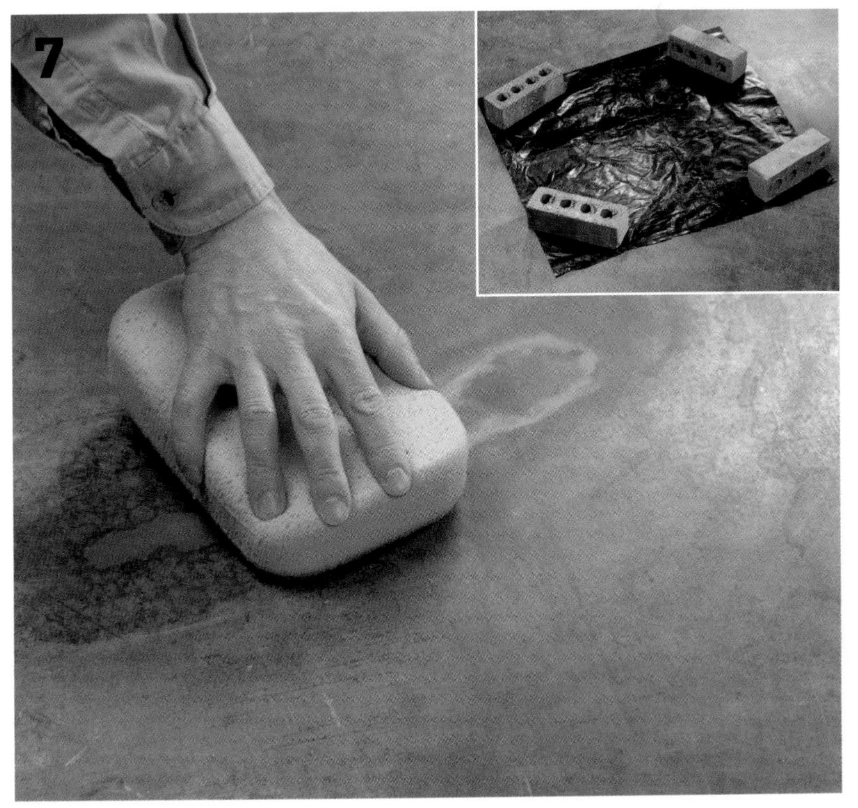

Use a slightly damp sponge to smooth out the remaining imperfections and clean off the edges. This actually works wonders and is a great way to compensate for being inexperienced at doing finish work with the trowel. (Inset) Cover with plastic. Some patch materials suggest that you cover the finished patch with a piece of plastic for a few days. This keeps the moisture from evaporating and gives the cement more time to cure. Make sure to weigh down the edges of the plastic. After the compound is cured, you can use the edge of the trowel to scrape the patched areas smooth, if necessary.

Glossary

Adhesive — Bonding agent used to adhere the floor covering to the underlayment. Adhesives are also available for installing a floor covering on nonporous surfaces, such as sheet vinyl.

Air bubbles — Pockets of air that get trapped under resilient sheet flooring, an indication that the adhesive has failed.

American National Standards Institute (ANSI) — A standards-making organization that rates tile for water permeability.

Baseboard — Strip of wood molding, available in various designs and thicknesses, applied at the bottom of the wall to cover the gap between the floor covering and the wall.

Baseboard shoe — A narrow piece of molding, often quarter round, attached to the bottom of baseboard to hide gaps between the floor covering and baseboard, and to add a decorative edge.

Baseboard tile — Baseboard-shaped tile used instead of wood baseboards. Used in conjunction with tile floors.

Batter boards — Temporary stake structures used for positioning layout strings for outdoor floors.

Beam — Any horizontal framing member such as a joist or header.

Berber carpet — Looped pile running in parallel lines. Berber carpet has the same color throughout the fibers.

Blindnail — Driving nails at an angle through the tongues of hardwood flooring so the next piece of flooring will cover the nail.

Border — Wood or tile of a different color or style than the main floor covering that's installed along the edge of a floor or around a design to add a decorative element.

Building code — A set of building regulations and ordinances regulating the way a house can be built or remodeled. Most building codes are controlled by a local municipality.

Building permit — Permit obtained from the local building department allowing you to remodel your home.

Carpet bar — A metal bar providing a transition between carpet and another floor covering that's at the same height or lower than the bottom of carpet.

Cementboard — Underlayment used for ceramic tile and some hardwood installations. Cementboard is the best underlayment in areas likely to get wet.

Chalk line — The line left by chalk, usually blue or red, after the chalk string is pulled tight between two points and snapped against the floor.

Clear finish — A wood finish that allows the wood grain to be seen without discoloring the wood.

Coefficient of friction— The measure of a tile's slip resistance. Tiles with high numbers are more slip resistant.

Common nail — A heavy-shaft nail used primarily for framing work, available from 2d to 60d.

Cross bridging — Diagonal braces installed between joists to keep them from moving and to keep floors from squeaking. Cross bridging can be wood or metal.

Crosscut — Cutting a piece of wood perpendicular to the wood grain.

Cut-pile carpet — Individual carpet fibers woven tightly together. The fibers are colored on the outside, but not on the inside.

Door casing — Wood molding and trim placed around a door opening to give it a finished look.

Dry-fit — Installing tile without mortar to test the layout.

Dry mix — Packaged mix, usually sold in bags, that can be combined with water to form mortar.

Embossing leveler — A mortarlike substance used to prepare resilient flooring or ceramic tile for use as an underlayment.

Endnail — Driving nails through the face of one board into the end of another one.

Engineered flooring — Flooring that's manufactured to look like solid hardwood, but is easier to install, less expensive, and more resistant to wear. Engineered flooring is available in strips or planks.

Expansion joint — A joint in a tile layout filled with a flexible material, such as caulk, instead of grout. The expansion joint allows the tile to shift without cracking.

Facenail — Driving nails into the face of tongue-and-groove flooring.

Fiber/cementboard underlayment — A thin, high-density underlayment used under ceramic tile and resilient flooring where floor height is a concern.

Field tile — Tile that's not part of a design or border.

Finish nail — A nail with a small, dimpled head, used for fastening wood trim and other detailed work.

Flagstone — Quarried stone cut into slabs usually less than 3" thick, used for outdoor floors.

Floating floor — Wood or laminate floor covering that rests on a thin foam padding and is not fastened or bonded to the subfloor or underlayment.

Floor board — A strip or plank in a wood floor.

Floor tile — Any type of tile designated for use on floors.

Floor-warming systems — A system of heating elements installed directly under the floor covering. Floor-warming systems provide supplemental radiant heat to warm up a floor.

Framing member — A common term for a single structural element of a construction framework, such as a stud, joist, truss, or beam.

Full-spread vinyl — Sheet vinyl with a felt-paper backing that is secured to the underlayment with adhesive.

Grout — A dry powder, usually cement-based, that is mixed with water and pressed into the joints between tiles. Grout also comes with latex or acrylic additive for greater adhesion and impermeability.

Horizontal span — The horizontal distance a stairway covers.

Isolation membrane — A flexible material installed in sheets or troweled onto an unstable or damaged base floor or subfloor before installing tile. Isolation membrane prevents shifts in the base from damaging the tile above.

Jamb — The top and side pieces that make up the finished frame of a door opening.

Joists — The framing members that support the floor.

Latex patching compound — Compound used to fill cracks and chips in old underlayment and to cover screw or nail heads and seams in new underlayment.

Level — A line or plane that is parallel to the surface of still water.

Longstrip flooring — Wood flooring that has multiple strips, usually three, fastened together to form a single plank.

Medallion — Wood or tile design placed in a floor as a decoration.

Miter cut — An angle cut in the end of a piece of flooring or molding.

Molding — Decorative strips of wood installed along walls and floors.

Natural stone tile — Tile cut from marble, slate, granite, or other natural stone.

On center — The distance from the center of one framing member to the center of the next.

Perimeter-bond vinyl — Sheet vinyl with a PVC backing that is placed directly on underlayment and secured by adhesives along the edges and seams.

Planks — Wood or laminate flooring that is 4" or more wide.

Plywood — A common underlayment for resilient and ceramic tile installations.

Portland cement — A combination of silica, lime, iron, and alumina that has been heated, cooled, and pulverized to form a fine powder from which mortar products are made.

PVC — Acronym for polyvinyl chloride. PVC is a rigid plastic material that is highly resistant to heat and chemicals.

Reducers — Strips of wood that provide a transition from a hardwood floor to an adjacent floor of lower height.

Reference lines — Lines marked on the subfloor to guide the placement of the floor covering.

Rip — Cutting a piece of wood parallel to the grain.

Rise — The height of a step in a stairway.

Riser — A board attached to the front of a step between treads in a stairway.

Run — The length of a step in a stairway.

Sealants — Product used to protect non- and semi-vitreous tile from stains and water damage. Sealants are also used to protect grout.

Sheet vinyl — Flooring material made from vinyl and other plastics in the form of sheets that are 6 ft. or 12 ft. wide and approximately 1⁄8" thick.

Sistering — Fastening a new floor joist to a damaged floor joist for additional strength.

Sleepers — Boards placed over a concrete floor and used to support the subflooring of a new floor.

Spacers — Plastic lugs inserted between tiles to help maintain uniform installation during installation.

Stain — Water-based or oil-based agent used to penetrate and change the color of a wood floor.

Strips — Wood or laminate flooring that is less than 4" wide.

Subfloor — The surface, usually made of plywood, attached to the floor joists.

Tack cloth — Lint free cloth, usually cheese cloth, used to clean floors and wipe away dust. Tack cloth is treated with a resin to make it sticky.

Tackless strips — Strips of wood nailed around the perimeter of a room. The teeth of the strips hold carpet in place.

Threshold — The area in a doorway where two floor coverings meet.

Toenail — Driving a nail at a 45° through the side of one board into the face of another one.

Tongue-and-groove flooring — Wood or laminate floor coverings that have a tongue and a groove in each individual piece. The flooring is assembled by placing the tongue and groove joints together.

Underlayment — Material placed on top of the subfloor, such as plywood, fiber/cementboard, cementboard, and isolation membrane.

Vapor barrier — Plastic sheeting used as a barrier to keep water from a concrete floor from penetrating the floor covering installed over it.

Waterproofing membrane — A flexible, waterproof material installed in sheets or brushed on to protect the subfloor from water damage.

Resources

Armstrong World Industries
717-397-0611
www.armstrong.com
p. 4, 26 (top), 87, 108, 120, 205

Ceramic Tiles of Italy
www.italiatiles.com
p. 19 (top), 19 (lower left), 21 (top), 22 (lower right), 23 (lower), 25 (top and lower), 27 (lower), 28 (lower), 29 (top), 32 (lower), 33 (top left and right), 35 (top left and lower)

Crossville Porcelain Stone
931-484-2110
www.crossvilleceramics.com
p. 160

Eco Friendly
866-250-3273
www.ecofriendlyflooring.com
p. 36 (lower), 38 (lower), 39 (top and lower)

FLOR
Inspired modular floor covering.
866-281-3567
www.flor.com
p. 33 (lower), 38 (top)

Hakatai Enterprises, Inc.
888-667-2429
www.hakatai.com
p. 30 (top)

HomerWood
Hardwood Flooring
814-827-3855
www.homerwood.com
p. 18 (top), 29 (lower), 35 (top right)

IKEA Home Furnishings
610-834-0180
www.Ikea-USA.com
p. 26 (lower left)

Kentucky Wood Floors
502-451-6024
www.kentuckywood.com
p. 104

Marmoleum
866-Marmoleum
www.themarmoleumstore.com
p. 20 (lower), 22 (lower left), 23 (top), 28 (top), 34, 37 (lower)

MIRAGE Prefinished Hardwood Floors
800-463-1303
www.miragefloors.com
p. 17 (top left), 20 (top), 90

Oshkosh Designs
877-582-9977
www.oshkoshdesigns.com
p. 96 (top)

Plyboo
866-835-9859
www.plyboo.com
p. 36 (top), 37 (top)

Room and Board
800-301-9720
www.RoomAndBoard.com
p. 24 (top), 30 (lower), 31 (lower), 32 (top), 224

Teragren Fine Bamboo Flooring, Panels & Veneer
800-929-6333
www.teragren.com
p. 22 (top), 131, 132, 133, 134, 135

2nd Wind Exercise Equipment
952-544-5249
www.2ndwind.net
7585 Equitable Drive
Eden Prairie, MN 55344
p. 116 / treadmill available at 2nd Wind Exercise Equipment

Photo Credits

p. 4 / photo courtesy of Armstrong
p. 16 / photo © Branko Miokovic (istock)
p. 17 (top left) / photo courtesy of MIRAGE Prefinished Hardwood Floors
p. 18 (top) / photo courtesy of HomerWood
p. 18 (lower) / photo © Brand X Pictures / Alamy
p. 19 (top) / photo courtesy of Ceramic Tiles of Italy
p. 19 (lower left) / photo courtesy of Ceramic Tiles of Italy
p. 20 (top) / photo courtesy of MIRAGE Prefinished Hardwood Floors
p. 20 (lower) / photo courtesy of Marmoleum by Forbo Linoleum
p. 21 (top) / photo courtesy of Ceramic Tiles of Italy
p. 21 (lower) / photo © Inside / Beateworks.com
p. 22 (top) photo courtesy of Teragren
p. 22 (lower left) / photo courtesy of Marmoleum
p. 22 (lower right) / photo courtesy of Ceramic Tiles of Italy
p. 23 (top) / photo courtesy of Marmoleum
p. 23 (lower) / photo courtesy of Ceramic Tiles of Italy
p. 24 (top) / photo courtesy of Room and Board
p. 24 (lower) / photo Richard Hoffkins (istock)
p. 25 (top) / photo courtesy of Ceramic Tiles of Italy
p. 25 (lower) / photo courtesy of Ceramic Tiles of Italy
p. 26 (top) / photo courtesy of Armstrong
p. 26 (lower left) / photo courtesy of Ikea
p. 26 (lower right) / photo Dar Yang Yan (istock)
p. 27 (top) / photo Christopher Hudson (istock)
p. 27 (lower) / photo courtesy of Ceramic Tiles of Italy
p. 28 (top) / photo courtesy of Marmoleum
p. 28 (lower) / photo courtesy of Ceramic Tiles of Italy
p. 29 (top) / photo courtesy of Ceramic Tiles of Italy
p. 29 (lower) / photo courtesy of HomerWood
p. 30 (top) / photo courtesy of Hakatai
p. 30 (lower) / photo courtesy of Room and Board
p. 31 (top) / photo Teun van den Dries (istock)
p. 31 (lower) / photo courtesy of Room and Board
p. 32 (top) / photo courtesy of Room and Board
p. 32 (lower) / photo courtesy of Ceramic Tiles of Italy
p. 33 (top left) / photo courtesy of Ceramic Tiles of Italy
p. 33 (top right) / photo courtesy of Ceramic Tiles of Italy
p. 33 (lower) / photo courtesy of FLOR
p. 34 / photo courtesy of Marmoleum
p. 35 (top left) / photo courtesy of Ceramic Tiles of Italy
p. 35 (top right) / photo courtesy of HomerWood
p. 35 (lower) / photo courtesy of Ceramic Tiles of Italy
p. 36 (top) / photo courtesy Plyboo, Woody Harrelson Eco Suite
p. 36 (lower) / photo courtesy of Eco Friendly
p. 37 (top) / photo Benny Chan Photography, courtesy of Plyboo
p. 37 (lower) / photo courtesy of Marmoleum
p. 38 (top) / photo courtesy of FLOR
p. 38 (lower) / photo courtesy of Eco Friendly
p. 39 (top and lower) / photo courtesy of Eco Friendly
p. 87 / photo courtesy of Armstrong
p. 90 / photo courtesy of MIRAGE Prefinished Hardwood Floors
p. 96 / photo courtesy of Oshkosh Designs
p. 104 / photo courtesy of Kentucky Wood Floors
p. 108 / photo courtesy of Armstrong
p. 120 / photo courtesy of Armstrong
p. 131, 132, 133, 134, 135 / photos by Steve Galvin (Cpi) / Bamboo courtesy of Teragren
p. 160 / photo courtesy of Crossville Porcelain Stone
p. 166 / photo courtesy of Mohawk Industries
p. 167 (top) / photo © Getty Images / Ryan McVay
p. 167 (lower) / photo © Getty Images / Janis Christie
p. 198 / photo M. Eric Honeycutt (istock)
p. 205 / photo courtesy of Armstrong
p. 224 / photo courtesy of Room and Board

Conversion Charts

Lumber Dimensions

Nominal - U.S.	Actual - U.S. (in inches)	Metric
1 × 2	¾ × 1½	19 × 38 mm
1 × 3	¾ × 2½	19 × 64 mm
1 × 4	¾ × 3½	19 × 89 mm
1 × 5	¾ × 4½	19 × 114 mm
1 × 6	¾ × 5½	19 × 140 mm
1 × 7	¾ × 6¼	19 × 159 mm
1 × 8	¾ × 7¼	19 × 184 mm
1 × 10	¾ × 9¼	19 × 235 mm
1 × 12	¾ × 11¼	19 × 286 mm
1¼ × 4	1 × 3½	25 × 89 mm
1¼ × 6	1 × 5½	25 × 140 mm
1¼ × 8	1 × 7¼	25 × 184 mm
1¼ × 10	1 × 9¼	25 × 235 mm
1¼ × 12	1 × 11¼	25 × 286 mm

Nominal - U.S.	Actual - U.S. (in inches)	Metric
1½ × 4	1¼ × 3½	32 × 89 mm
1½ × 6	1¼ × 5½	32 × 140 mm
1½ × 8	1¼ × 7¼	32 × 184 mm
1½ × 10	1¼ × 9¼	32 × 235 mm
1½ × 12	1¼ × 11¼	32 × 286 mm
2 × 4	1½ × 3½	38 × 89 mm
2 × 6	1½ × 5½	38 × 140 mm
2 × 8	1½ × 7¼	38 × 184 mm
2 × 10	1½ × 9¼	38 × 235 mm
2 × 12	1½ × 11¼	38 × 286 mm
3 × 6	2½ × 5½	64 × 140 mm
4 × 4	3½ × 3½	89 × 89 mm
4 × 6	3½ × 5½	89 × 140 mm

Metric Conversions

To Convert:	To:	Multiply by:
Inches	Millimeters	25.4
Inches	Centimeters	25.4
Feet	Meters	0.305
Yards	Meters	0.914
Square inches	Square centimeters	6.45
Square feet	Square meters	0.093
Square yards	Square meters	0.836
Ounces	Milliliters	30.0
Pints (U.S.)	Liters	0.473 (Imp. 0.568)
Quarts (U.S.)	Liters	0.946 (Imp. 1.136)
Gallons (U.S.)	Liters	3.785 (Imp. 4.546)
Ounces	Grams	28.4
Pounds	Kilograms	0.454

To Convert:	To:	Multiply by:
Millimeters	Inches	0.039
Centimeters	Inches	0.394
Meters	Feet	3.28
Meters	Yards	1.09
Square centimeters	Square inches	0.155
Square meters	Square feet	10.8
Square meters	Square yards	1.2
Milliliters	Ounces	.033
Liters	Pints (U.S.)	2.114 (Imp. 1.76)
Liters	Quarts (U.S.)	1.057 (Imp. 0.88)
Liters	Gallons (U.S.)	0.264 (Imp. 0.22)
Grams	Ounces	0.035
Kilograms	Pounds	2.2

Counterbore, Shank & Pilot Hole Diameters

Screw Size	Counterbore Diameter for Screw Head (in inches)	Clearance Hole for Screw Shank (in inches)	Pilot Hole Diameter	
			Hard Wood (in inches)	Soft Wood (in inches)
#1	.146 (9/64)	5/64	3/64	1/32
#2	1/4	3/32	3/64	1/32
#3	1/4	7/64	1/16	3/64
#4	1/4	1/8	1/16	3/64
#5	1/4	1/8	5/64	1/16
#6	5/16	9/64	3/32	5/64
#7	5/16	5/32	3/32	5/64
#8	3/8	11/64	1/8	3/32
#9	3/8	11/64	1/8	3/32
#10	3/8	3/16	1/8	7/64
#11	1/2	3/16	5/32	9/64
#12	1/2	7/32	9/64	1/8

Index